Integrated Circuits and Systems

Series Editor
Anantha P. Chandrakasan

For further volumes:
http://www.springer.com/series/7236

Eugenio Cantatore
Editor

Applications of Organic and Printed Electronics

A Technology-Enabled Revolution

Springer

Editor
Eugenio Cantatore
Department of Electrical Engineering
Eindhoven University of Technology
Eindhoven
Netherlands

ISSN 1558-9412
ISBN 978-1-4614-3159-6 ISBN 978-1-4614-3160-2 (eBook)
DOI 10.1007/978-1-4614-3160-2
Springer Boston Heidelberg New York Dordrecht London

Library of Congress Control Number: 2012944381

Printed on acid-free paper

Springer is part of Springer Science+Business Media (www.springer.com)

Preface

The Disruptive Potential of Low-Cost, Low-Temperature Technologies for Electronics

Electronics, and more specifically integrated circuits (IC), have dramatically changed our lives and the way we interact with the world. Following the so-called Moore's law [1], IC complexity is growing exponentially since 40 years, and this trend is predicted to continue at least for the coming 15 years [2]. The abundance of electronic functions at affordable cost has enabled a wealth of applications where the main IC strengths, namely computational speed and memory capacity, are well exploited: PCs, portable devices, game consoles, smart phones and alike. The commercial success of integrated electronics is based on a symbiotic development of technology and applications, where technical progress and economic growth nurture each other. This process requires lots of time and effort: first IC patents where filed in 1949 [3], but it is only in 1971 that the first commercially available microprocessor (Intel 4004), one of the most far-reaching application of ICs, gained the market; and PCs became popular only in the second half of the eighties.

The main strength of integrated electronics is in the low-cost-per-function enabled by an ever growing miniaturization: mono-crystalline silicon real estate is very expensive, but the number of transistors that can be integrated per area grows according to Moore's law, bringing down the cost to realize a given function.

Since the second half of the seventies, a completely different electronic paradigm, the so-called large-area electronics, has been developing. In this field the major aim is to decrease the cost per area (instead of the cost per function), enabling large surfaces covered with electronic devices. The main application of this kind of technology, typically based on amorphous or polycrystalline silicon transistors, is in active-matrix addressing of flat displays. The success of this technology has become evident in the last decade, when flat-panel LCD displays have swiftly replaced traditional cathode ray tubes in television sets.

Amorphous and polycrystalline silicon technology typically require high-temperature vacuum-based processing, with the consequence that glass substrates are used and that the technology throughput is limited. In the nineties a new technology approach has been proposed, based on materials that enable low-temperature processing and the use of very high throughput patterning technologies, borrowed from the graphic printing field: organic and printed electronics were born.

The word "organic electronics", which I personally started using in 2000 [4] together with many colleagues, designates electronics manufactured using functional carbon-based materials, typically semiconductors, like pentacene, P3HT, PCBM, PTAA and many others. There are several reasons for this choice:

- Organic materials can form functional films when processed from solutions, paving the way to manufacturing processes with a reduced number of vacuum steps (which are typically expensive and cumbersome to scale to large areas), and thus enabling potentially very low-cost large-area electronics;
- Organic materials are processed at low temperature (typically below 200 °C), enabling the use of inexpensive and flexible plastic foils as substrates and paving the way to flexible electronics;
- Organic chemistry is intrinsically very rich, enabling the exploration of a limitless library of materials having very diverse electrical, optical, rheological and chemical properties;
- Together with the chemical variety, a large spectrum of physically different devices based on organic materials is possible and has been developed in the years, the most well-known being organic light emitting diodes (OLEDs) [5], organic thin-film transistors (OTFTs) [6, 7], organic photovoltaics (OPVs) [8], organic sensors [9], organic memories [10, 11], and organic MEMs [12][1].

Together with these strengths, functional organic materials and organic electronics present a number of drawbacks:

- Organic semiconductors have a relatively poor mobility, with peak values for single-crystal materials in the range of 10 cm^2/Vs [13], and typical values in solution-processed films of about 1 cm^2/Vs at the state of the art. Under this point of view, other materials suitable for low-temperature and large-area processing, like metal-oxide semiconductors and carbon nanotubes, may offer an advantage compared to organic semiconductors.
- Organic semiconductors (especially n-type) are sensitive to oxygen, moisture and other environmental aggressors, so that for long time organic electronic devices have had poor shelf and operational lifetime. Organic materials are also sensitive to bias stress, which tends to affect operational lifetime. Recent improvements in the materials, their formulation and encapsulation, however, show that instabilities should not be a show-stopper for commercialization (see for instance Sect. 2.3 in Chap. 2 and Sect. 4.4 in Chap. 4);

[1] In this section a few early and significant papers have been selected as references.

- Organic semiconductors are difficult to dope in situ with highly controlled dopant concentrations as a process equivalent of the ion implantation doping used in silicon has still not been developed for organic materials. This makes difficult to manage key parameters like transistor threshold voltages and injection barriers at the contacts.

Many more details on the state of the art and roadmaps of organic electronics are given in Chap. 1 and in the other chapters of this book.

The capability to deposit organic materials from solution makes possible to pattern functional materials using methods adapted from graphic printing, like inkjet, gravure, slot coating and many others. This leads to the concept of "printed electronics". The main strength of this approach is the high throughput that characterizes printing production processes, which means that printing has the potential to make possible very inexpensive large-area electronics, and thus to enable applications of electronics unthinkable till now. Moreover, printing is an additive process, thus only the functional materials that are needed are effectively used, contrary to the traditional lithography-based subtractive approach. This has the potential to decrease material usage and thus further bring down the costs. Detailed information on printing electronics is available especially in Chaps. 1, 2 and 6 of this book.

The strengths of printing are paired with the challenges that this technology faces: it is namely difficult and expensive to develop a new electronic technology using an approach that in a few minutes can generate rolls covered with hundreds of meters of electronics to be characterized and optimized. Uniformity, performance and yield are daunting tasks to be solved for future printed electronics applications.

The potential low cost, the compatibility with large flexible substrates and the wealth of devices that characterize organic and printed electronics will make possible applications that go far beyond the well-known displays made with conventional large-area silicon electronics. Organic and printed electronics can enable a true revolution in the applications of electronics: this is the view that brought me, together with a large number of colleagues, to write this book. The volume offers to the reader an extensive overview of the different devices enabled by organic electronics, and reviews a large variety of applications that are developing and can be foreseen for the future.

Chapter 1, written by Tampere University, the Organic Electronic Association (OA-E) and PolyIC, offers a complete *Roadmap for Organic and Printed Electronics* spanning till the end of this decade. It is an ideal starting point to understand the complex application scenarios and the likely developments in this rapidly growing technology domain.

In Chap. 2 by Konarka, Cyprus University of Technology and Friedrich-Alexander-University, are discussed *Organic Photovoltaics*, with great emphasis on the use of printing processes for their manufacturing. A wide overview of the printing processes for organic electronics is given, together with the state of the art of their application to solar cells. Photovoltaic cells do not need fine patterning of the structures in the plane of the device, and are thus an ideal candidate to exploit the high throughput of printing processes. This chapter is an excellent reading for the person willing to understand more about printing electronics. A roadmap for organic solar cells concludes this contribution.

In the third and fourth chapter light emitting diodes (OLED), the most advanced organic electronic devices available at the moment, are discussed. Chapter 3, written by Kyung Hee University and Samsung, gives a detailed overview of *OLED Displays*, a booming application that has reached the market since some years already, and is rapidly growing to become the standard emissive technology for flat displays. This section informs the reader about the different types of OLED pixels in commercial use and in development, and gives insight into the most relevant display and backplane issues.

Chapter 4, by Philips, gives a nice overview of *OLED for Lighting* applications. The section begins with an insightful description of the materials, physics, architecture and benchmarking of OLED lighting devices, to continue with an overview of fabrication methods, reliability and commercial applications.

Chapter 5 by University of Tokyo gives an interesting vision for future organic electronics: it will complement silicon ICs to create new applications enabling unprecedented ways of interaction between electronics and people. In this vision are included a variety of different organic devices (TFTs, sensors and actuators) providing a stimulating view on how different types of organic electronics can be integrated to enable revolutionary applications.

The sixth and seventh chapter deal with organic TFTs. Chapter 6 focuses on applications of *Printed Organic TFTs*. This section, written by PolyIC, describes the devices and technology needed to print transistors and circuits, the characteristics of printed TFTs, and what this revolutionary technology can mean in terms of applications (*RFIDs and Smart Objects*). Chapter 7 by IMEC, KUL, KHL, TNO and Polymer Vision focuses on the application of *Organic TFTs* to low-cost *RFIDs*. This section explains how organic RFIDs are developing towards becoming fully-compliant to existing standards for RFIDs based on silicon IC technology. Compatibility with standards would mean that the same infrastructure can be shared between silicon and organic RFIDs, enabling a seamless transition between the two technologies and an easy market uptake. This does not mean, however, that silicon and organic should serve the same markets: the characteristics of printed electronics lend themselves naturally to the dream of enabling item-level identification of retail items, which is still out of reach for silicon RFIDs, due to the high costs and cumbersome integration of silicon ICs with the items to be identified.

Chapter 8, contributed by University of California Berkeley, reviews the state of the art of *Chemical Sensors* based on organic electronic devices and demonstrates the specific competitive advantage that these sensors have, namely the ease of creating matrices of sensing elements with different sensitivity to diverse analytes, thus enabling the extraction of unique analyte signatures and greatly improving both specificity and versatility of use.

This book can be read at different levels of insight by beginners as well as by experts in the field, and is specifically conceived to address a wide range of people with technical and scientific background. I am deeply grateful to all contributors: I hope you will appreciate their effort and I wish you a pleasant and fruitful reading.

Eindhoven, The Netherlands, January 2012 Eugenio Cantatore

References

1. Moore GE (2003) No exponential is forever: but "forever" can be delayed! In: ISSCC 2003 digest of technical papers, pp 20–23
2. ITRS Roadmap (2011) Available at http://www.itrs.net/Links/2011ITRS/Home2011.htm
3. Jacobi W (1949) Halbleiterverstärker, Patent DE833366, 15 April 1949
4. Cantatore E (2001) State of the art electronic devices based on organic materials. In: Proceedings of the 31st European solid-state device research conference (ESSDERC), pp 25–34
5. Tang CW, VanSlyke SA (1987) Organic electroluminescent diodes. Appl Phys Lett 51:913
6. Koezuka H, Tsumura A, Ando T (1987) Field-effect transistor with polythiophene thin film. Synth Met 18:699–704
7. Brown AR, Pomp A, Hart CM, de Leeuw DM (1995) Logic gates made from polymer transistors and their use in ring oscillators. Science 270(5238):972–974
8. Sariciftci NS, Smilowitz L, Heeger AJ, Wudl F (1992) Photoinduced electron-transfer from a conducting polymer to buckminsterfullerene. Science 258(5087):1474–1476
9. Torsi L, Dodabalapur A, Sabbatini L, Zambonin PG (2000) Multi-parameter gas sensors based on organic thin-film-transistors. Sens Actuators B 67:312
10. Reed MA, Chen J, Rawlett AM, Price DW, Tour JM (2001) Molecular random access memory cell. App Phys Lett 78(23):3735–3737
11. Ouyang JY, Chu CW, Szmanda CR, Ma LP, Yang Y (2004) Programmable polymer thin film and non-volatile memory device. Nat Mater 3(12):918–922
12. Sekitani T, Takamiya M, Noguchi Y, Nakano S, Kato Y, Hizu K, Kawaguchi H, Sakurai T, Someya T (2006) A large-area flexible wireless power transmission sheet using printed plastic MEMS switches and organic field-effect transistors. In: IEEE int. electron devices meeting (IEDM), pp 287–290
13. Jurchescu OD, Popinciuc M, van Wees BJ, Palstra TTM (2007) Interface-controlled, high-mobility organic transistors. Adv Mater 19:688–692

Contents

Chapter 1
OE-A Roadmap for Organic and Printed Electronics

Donald Lupo, Wolfgang Clemens, Sven Breitung and Klaus Hecker

Abstract The roadmap for organic and printed electronics is a key activity of the OE-A, the industrial organisation for the young organic, printed and large area electronics industry. Organic electronics is a platform technology that enables multiple applications, which vary widely in their specifications. Since the technology is still in its early stage—and is in the transition from lab-scale and prototype activities to production—it is important to develop a common opinion about what kind of products, processes and materials will be available and when. This chapter is based on the third version of the OE-A Roadmap for organic and printed electronics, developed as a joint activity by key teams of experts in 9 applications and 3 technology areas, informed by further discussions with other OE-A members during association meetings. The resulting roadmap is a synthesis of these results representing common perspectives of the different OE-A forums. Through comparison of expected product needs in the application areas with the expected technology development paths, potential roadblocks or "red brick walls" such as resolution, registration and complementary circuitry are identified.

D. Lupo (✉)
Department of Electronics, Tampere University of Technology,
PO Box 692, 33101 Tampere, Finland
e-mail: donald.lupo@tut.fi

W. Clemens
PolyIC GmbH and Co.KG, Tucherstrasse. 2, 90763 Fürth, Germany
e-mail: wolfgang.clemens@polyic.com

S. Breitung · K. Hecker
OE-A (Organic Electronics Association), c/o VDMA, Lyoner Street 18,
60538 Frankfurt am Main, Germany
e-mail: sven.breitung@vdma.org

K. Hecker
e-mail: klaus.hecker@vdma.org

E. Cantatore (ed.), *Applications of Organic and Printed Electronics*,
Integrated Circuits and Systems, DOI: 10.1007/978-1-4614-3160-2_1,

Keywords Organic electronics · Printed electronics · Roadmap OE-A applications · Red brick walls · Organic electronics association

1.1 Introduction

Organic and printed electronics is based on the combination of new materials and cost-effective, large area production processes that open up new fields of application. Thinness, light weight, flexibility and environmental sustainability are key advantages of organic electronics. Organic electronics also enables a wide range of electrical components that can be produced and directly integrated in low cost reel-to-reel processes.

Intelligent packaging, low cost RFID (radio-frequency identification) transponders, rollable displays, flexible solar cells, disposable diagnostic devices or games, and printed batteries are just a few examples of promising fields of application for organic electronics based on new, large scale processable, electrically conductive and semi-conducting materials.

The following pages present a short overview of organic electronics applications, technologies and devices, as well as a discussion of the different technology levels that can be used in manufacturing organic electronic products, based on the third edition of the roadmap developed by the OE-A. Since the second edition we have added further applications that we expect to play a key role in the commercialization of this emerging technology and taken account of the exciting technical progress made recently.

In the applications section which follows, the market entry on larger scales for the various applications is forecasted. The key application and technology parameters relating to these applications and the principle challenges (so-called red brick walls) to achieving these have been identified. In the subsequent technology section we summarise the projected development of relevant technologies and take account of recent progress in new materials and improved processes.

A White Paper explaining the current edition of the roadmap in more detail can be downloaded [1].

Organic electronics

Organic electronics is based on the combination of a new class of materials and large area, high volume deposition and patterning techniques. Often terms like printed, plastic, polymer, flexible, printable inorganic, large area or thin film electronics or abbreviations like OLAE or FOLAE (Flexible and/or Organic Large Area Electronics) are used, which essentially all mean the same thing: electronics beyond the classical integrated circuit approach. For simplicity we have used the term organic electronics in this roadmap, but keep in mind that we are using the term in this broader sense.

1.2 Applications

Organic and printed electronics is a platform technology that is based on organic conducting and semi-conducting as well as printable inorganic materials. It opens up new possibilities for applications and products. A number of key applications of organic and printed electronics have been chosen to demonstrate the needs from the application side, identify major challenges, cross check with the possibilities of the technology and to forecast a time frame for the market entry in large volumes.

Below, we continue to look at applications discussed in the second edition of the roadmap. i.e. organic **photovoltaic cells (OPV)**, printed **RFIDs**, organic **memories**, organic **sensors**, flexible **batteries** and **smart objects**. We also expand on the previous application area of organic thin film transistor (OTFT) display backplanes to look at **flexible displays**, and look at two new application areas, electroluminescence (EL) and organic LED **(OLED) based lighting** and **smart textiles**.

The growing list of applications reflects the complexity of the topic and the wide possible uses for organic electronics, and it is likely that the list will even grow in the future. The application fields and specifications cover a wide range, and although several parameters like accuracy of the patterning process or electrical conductivity of the materials are of central importance, the topic cannot be reduced to one single parameter at the time being, as is known from the famous Silicon Roadmap (Moore's law). Regardless, we will watch the trends and find out whether it will be possible to find an analogue to Moore's law for organic electronics.

The question whether there is one "killer application" for organic electronics cannot be answered at this moment. There are many different fields in which the advantages of organic electronics might result in the right product to become the killer application, but at this point, it is too early to define which one it is. Past experience with new technologies has shown that the predicted "killer applications" are frequently not the ones that really open up the largest markets. Therefore, one has to continue the work on the roadmap, as is planned, follow the actual trends and take account of new developments as they occur.

First organic electronic products reached the market in 2005/2006. OLED displays are not specifically covered as such in this version of the roadmap but are also based on organic semiconductors, and are starting to see substantial market penetration in recent years. Passive ID cards that are mass printed on paper and are used for ticketing or toys were presented in 2006 [2]. Flexible Lithium batteries—produced in a reel-to-reel process—have been available for several years and can be used for smart cards and other mobile consumer products [3]. Printed antennae are commonly used in (still Si-based) RFID tags. Large-area organic pressure sensors for applications such as retail logistics have also been introduced, as have printed electrodes for glucose test strips. Recently, first OPV [4, 5] and OLED lighting based products [6, 7] have become available and first user tests of smart cards with built-in displays for one-time password applications have been started.

Additional products, like glass-free high resolution e-readers or rollable displays with organic TFT backplanes, printed radio frequency tags and organic

Fig. 1.1 Bag with integrated OPV battery charger. *Source* Neubers

memories, have already been demonstrated technically and have recently approached the market. Within 2–4 years, it is expected that mass markets will be reached and that all the above mentioned applications, and several more, will be available in large volumes.

1.2.1 Applications Roadmap

Dye sensitised solar cell (DSSC) based **organic photovoltaic** products have been produced commercially since 2007 [8]. First polymer OPV products have been shipped, with increasing commercial availability, e.g. as flexible solar cells (see Fig. 1.1) for a battery charger for mobile phones. For the next few years OPV will primarily address consumer, outdoor recreational and initial off-grid markets, but as efficiency and lifetime improve the target is to move into building integrated PV (BIPV) and off-grid power generation mid-term and, in the long term, enter the on-grid power generation market. This will require significant technical progress in materials and processes to deliver high efficiency, highly stable products. In this book organic photovoltaics are further discussed in Chap 2.

Flexible displays are starting to enter the market, with roll to roll produced segmented electrophoretic price labels already being used in stores and rollable e-reader devices with OTFT backplanes (Fig. 1.2) and large area unbreakable OTFT based e-reader products test marketed 2011. Displays based on **electrophoretic** or **electrochromic** media or on **OLEDs** are currently getting a particularly large amount of attention, but displays based on **liquid crystals, electrowetting** etc. are also possible. Further in the future, both reflective and emissive colour displays and large area products like rollable OLED TVs or electronic wallpaper are anticipated. However, the move to colour, high resolution and OLEDs will require significant improvements in backplane patterning technology, display media and OTFT technology.

Fig. 1.2 Rollable electrophoretic display for e-readers and mobile phones. *Source* Polymer vision

Electroluminescent (EL) and OLED Lighting is an application that is new to the third edition of the roadmap. While OLEDs have been penetrating the display market for some time now, only recently have significant improvements in efficiency, lifetime and large area devices made OLED an important potential source of novel large-area, energy efficient solid-state lighting. EL signage and backlighting is already commercial, first OLED designer lamps (Fig. 1.3) are already available, and in the future OLED lighting will move from being a technology for design and decorative applications to technical lighting and general illumination; this will however, require both very high efficiency, colour purity and lifetime as well as development of processes, materials and architectures to cut production costs. Chap. 4 of this book further addresses OLED lighting and its applications.

Printed RFID (radio frequency identification) based on organic electronics showed significant technical progress since the last edition of the roadmap, with announcements of advances such as roll to roll printed high frequency (HF) tags with 1–4 bits, as well as first organic CMOS-like circuits [9], 128 bit transponders [10], and ultrahigh frequency (UHF) rectifiers [11], all based on organic semiconductors. In addition, there has been progress with alternative approaches such as chipless RFID concepts. Printed antennas are already common in conventional Si-based RFID products. A further approach for printed transponders is based on Si nanoparticles on stainless steel substrates. These approaches are not further taken into account in the current roadmap discussion, as this roadmap focuses on organic/printed chips on plastic substrates. The activities of printed RFID are targeting towards Electronic Product Code (EPC^{TM}) compatible tags in the long term (see Chap. 7), even though the general performance of printed RFID will be on a lower level compared to standard RFID tags for a long time. Simple printed

Fig. 1.3 OLED designer lamp. *Source* OSRAM Opto semiconductors

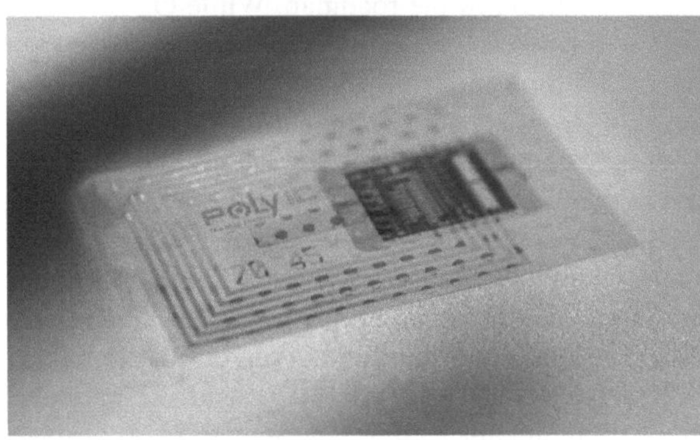

Fig. 1.4 Printed RFID tag. *Source* PolyIC

RFID tags (Fig. 1.4) were piloted already in 2007 and should be in general commercial use within the next few years. The future is expected to bring a trend to larger memory, and to UHF as well as HF tags. The expected applications range from brand protection into ticketing, identification, automation and logistics, as the technology advances. Despite some delays in market introduction of simple RF circuits, the rapid technical progress in the recent past makes us optimistic that more advanced products will actually be available within the next years. Keys to this progress will be mature high volume and low cost production processes, fast circuits, smaller dimensions and CMOS-like circuit development, as well as appropriate standards for organic RFID products. RFIDs are the main subject of Chaps. 6 and 7 of this book.

Fig. 1.5 Game cards with organic NV-RAM. *Source* Thin film electronics

Printed Memory devices have already been introduced to the market in the form of Read-only Memories (ROM) or Write Once Read Many (WORM) memories in ID or game cards. Recently reel to reel fabrication of printed rewritable non-volatile Random Access Memories (NV-RAM) was technically demonstrated [12], and first low-density polymer NV-RAM products are available on the market (Fig. 1.5). Future generations of printed memory products will see a trend to higher bit density, faster reading and writing, on-board readout and a trend to more NV-RAM, though ROM and WORM will remain important. Key technical issues to resolve in the future will include scaling of on-board readout electronics and memory cells.

Organic Sensor devices (Fig. 1.6) open up a variety of applications. The field has developed more rapidly than expected, with prototype temperature, chemical and pressure sensors already demonstrated. Temperature, pressure and photodiode sensors and sensor arrays will reach the market in the next few years. One trend will be from yes/no sensors to analog sensors able to give a quantitative readout. For example, potentiometric sensors for chemical analysis are already starting to become available in a yes/no configuration but analog versions will be available midterm. In the long term, combination of sensor devices into embedded systems including on-board (organic) circuitry and possibly on-board display-based readout is expected to enable intelligent sensor systems. This will require significant advances not only in the sensors themselves but also in the associated on-board circuitry, which will require high reproducibility, reliability, yield, etc. Sensors are further discussed in Chap. 8 of this book, while integration with circuits to enable intelligent sensor systems is addressed in Chaps 6 and 7

Thin and **flexible batteries** (Fig. 1.7) are already commercially available for discontinuous use, but there is room for improvement in price, capacity and ease of integration into some systems. Over the next few years a trend to commercial availability of cost-effective low capacity batteries, then higher capacity batteries for continuous use and finally batteries that can be directly printed into electronic

Fig. 1.6 Large-area organic based pressure sensor array. *Source* Plastic electronic

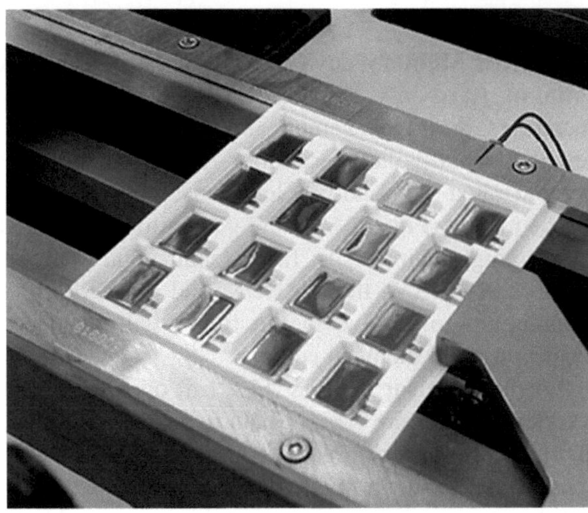

Fig. 1.7 Ultraslim primary batteries for mobile devices. *Source* VARTA microbattery

systems or packages is expected. Key areas for development will be optimisation of cost-effective production and encapsulation of Li based thin batteries.

A big advantage of organic electronics is the combination and simple integration of multiple electronics devices to create **smart objects**. As simple example, printed keypads, printed loudspeakers and smart cards incorporating thin film batteries and flexible displays (Fig. 1.8) have been shown [13]. In the future the trend will be towards inclusion of more different functionalities as well as more complex functionalities, moving from simple input devices, animated logos or smart cards to objects with full displays, intelligent tickets and sensors, games, and smart packages. The variety of smart objects will be limited only by the number of organic electronic technologies available and the creativity of product developers. One of the key issues to look at will be taking care of mechanical and electrical compatibility and connection between the different functions.

Another new application in the current roadmap is **smart textiles**, in which functionalities such as communication, displays, sensors, or thermal management

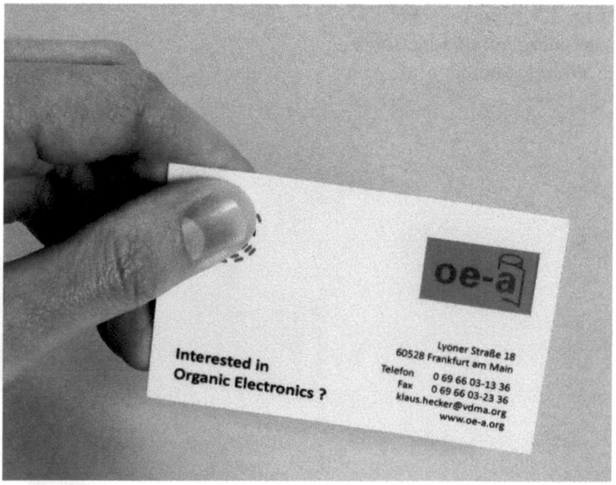

Fig. 1.8 Smart card with flexible battery and electrochromic display. *Source* OE-A

are integrated into fabric to enable wearable electronics (Fig. 1.9). First examples of integration of LEDs, optical fibers or electroluminescent elements into apparel are already starting to hit the market [14, 15]. Application areas range from sport, fashion, safety and health clothing to architecture, and over time the technology will become more complex, moving from simple sensors, keypads, light effects etc. in the short term to more complex systems incorporating functionalities like OPV, fuel cell and textile sensors in the future.

These application scenarios are summarized in the **OE-A roadmap for organic electronics applications** in Fig. 1.10. For each of the nine selected applications we show products that are expected to reach the market in the short (2009–2012) and medium term (2012–2017). We also give a forecast for the long term, from 2018 onward. Such a summary over many applications is by necessity not detailed; for each application area individual roadmaps have been prepared (see for example roadmaps for RFID and OPV in Figs. 1.11 and 1.12). Figure 1.10 is a high-level overview for the whole field of organic and printed electronics that has been distilled from the individual roadmaps.

This list of products reflects the ideas from today's point of view. Past experience of new technology shows us that we are most likely to be surprised by unexpected applications, and this will almost certainly happen in the exciting but nascent field of organic electronics. Therefore the technology and the market in this field will continuously be watched and the roadmap will be updated on a regular basis.

Significant progress has been made in the last several years and first generations of products have already been enabled. However, in order to fulfil the more demanding specifications of more complex future generations of products, further improvement of materials, process, design and equipment is necessary. In the next section we look at some of the main application parameters whose development will be key to enabling future product generations. After that we will look at the main technologies in organic electronics and discuss the key technology parameters underlying the application parameters.

Fig. 1.9 Sports jacket including smart functions. *Source* Francital

1.2.2 Key Application Parameters

The viability of each application or product will depend on fulfilment of a number of parameters that describe the complexity or performance of the product (application parameters). For the applications described above groups of specialists identified the most important application and technology parameters and requirements for different generations of products. Here we list only a small excerpt of the key application parameters that have been identified as relevant to several of the applications. The following list is in no particular order since the relevance of the different parameters varies for the diverse applications.

- **Complexity of the device**

The complexity of the circuit (e. g. number of transistors) as well as the number of different devices (e. g. circuit, power supply, switch, sensor, display) that are integrated have a crucial influence on reliability and production yield.

- **Operating frequency of the circuit**

With increasing complexity of the application (e.g. increasing memory capacity) higher switching speeds are necessary.

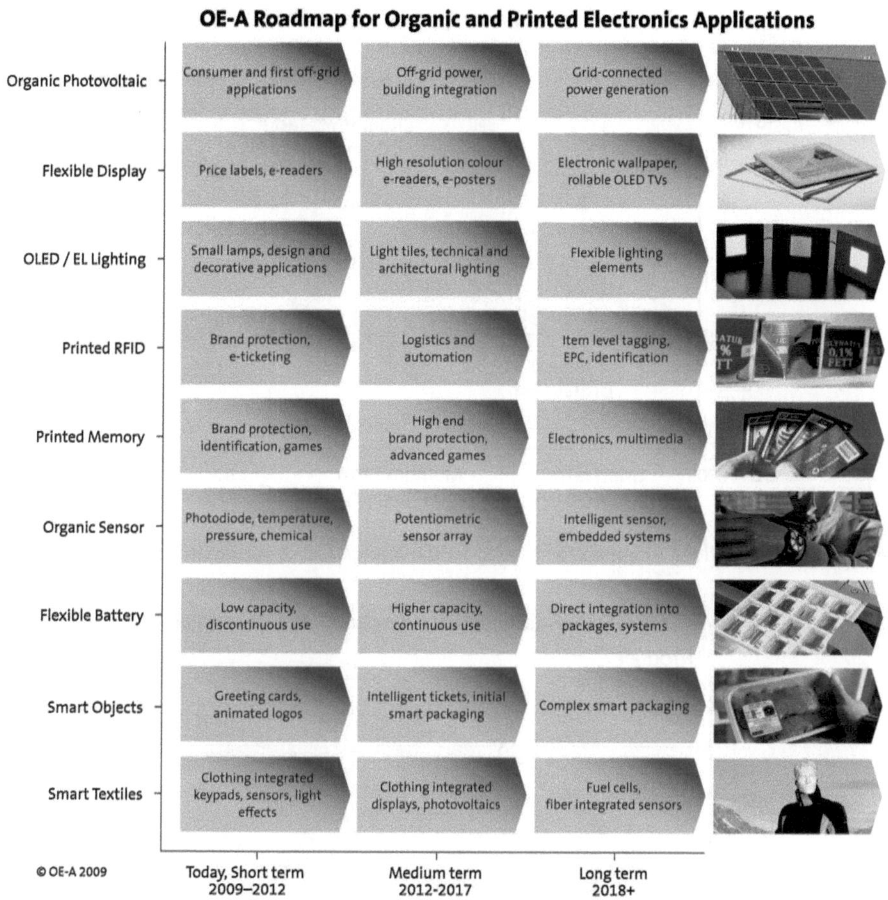

Fig. 1.10 OE-A Roadmap for organic electronics applications. Forecast for the market entry in large volumes (general availability) for the different applications. Source OE-A

- **Lifetime/stability/homogeneity**

Lifetime (shelf and operation), the environmental stability, stability against other materials and solvents, and homogeneity of the materials are issues due to the intrinsic properties of the materials used in organic and printed electronics.

- **Operating voltage**

For mobile devices powered by batteries, PV or radio frequency, it is essential to have low operating voltages (<10 V).

- **Efficiency**

The conversion efficiency of light to electricity or electricity to light is a key parameter for photo-voltaic cells and photodiodes or OLEDs, and power efficiency

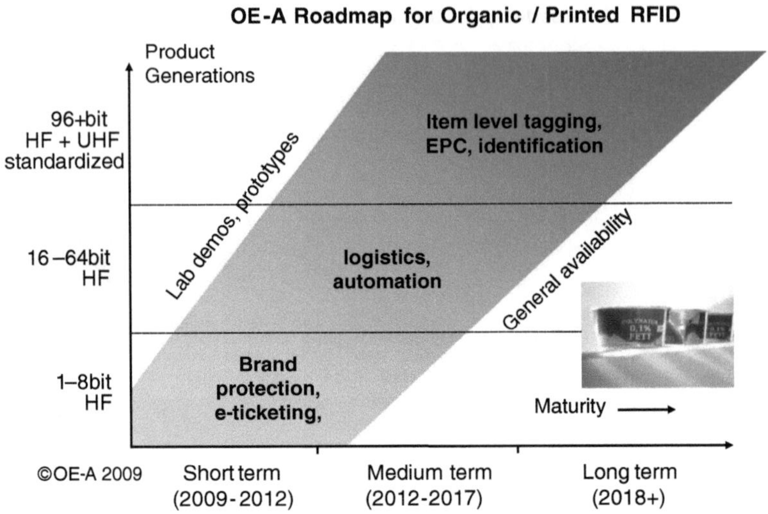

Fig. 1.11 Applications roadmap for printed RFID. *Source* OE-A image and *Source* PolyIC

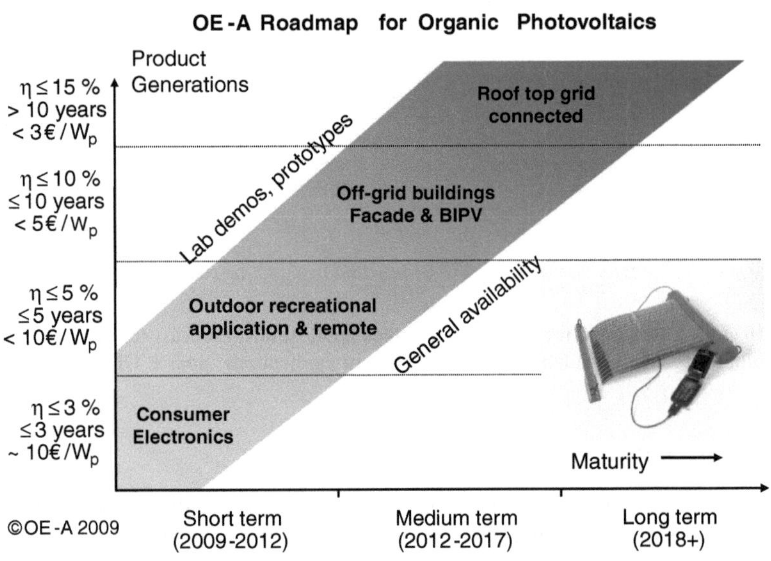

Fig. 1.12 Applications roadmap for OPV. *Source* OE-A image and *Source* Konarka

of circuitry is also important for many applications, especially those which are mobile and need to be light weight.

- **Cost**

Although most applications target new applications and markets rather than replacements, costs have to be low. For some applications, such as rollable displays, a cost premium over conventional rigid displays may be accepted, while for other applications, e.g. in packaging, low cost will be a major driving factor.

1.3 Technology

As we have mentioned before, we use the term organic electronics for brevity to refer to the field of electronics beyond classical silicon IC approaches, but include concepts such as large area or flexible circuits and printed inorganic materials. Although some classic **device** concepts are used, **materials,** including **substrates,** and **patterning processes** are very different from those used in the conventional electronic industry. In this section we review key materials, processes and devices for organic electronics and discuss the key technology parameters that are critical for development of future products. A more detailed description of the printing and other patterning processes, materials and devices can be found in an article in the 1st edition of the OE-A brochure, published in 2006 [16] and in Chap. 2 of this book.

1.3.1 Materials

Organic electronics rely on electrically active materials such as conductors, semiconductors, dielectrics, luminescent, electrochromic, electrophoretic or encapsulation materials. The materials have to be carefully chosen since process conditions and the interplay with other layers have a large influence on the performance of the device; for example the choice of appropriate dielectrics and of encapsulation materials can be critical for the performance and the stability of an organic electronic product. In this edition of the roadmap we have focused primarily on conducting and semiconducting materials, though in future editions we plan to include other classes of materials as well.

There are many approaches on the material side and the pros and cons of the different approaches—organic or inorganic, solution based or evaporated—are still under discussion. It is very likely that several approaches will be used in parallel.

Organic **conductors** such as PEDOT:PSS are starting to be widely used for electrodes in a variety of applications. Organic conductors can be highly transparent, and with recent progress in conductivity PEDOT:PSS is starting to become a realistic replacement for Indium tin Oxide (ITO) in some applications (Fig. 1.13). Inorganic materials like silver and other metals (e. g. as filled pastes or ultra-thin films) are also useful if still higher conductivity is needed.

Organic **Semiconductors** are used in numerous active devices and many of them are solution processable and can be printed. Figure 1.14 shows the structures of the organic conductor PEDOT:PSS, of the common polymer semiconductor

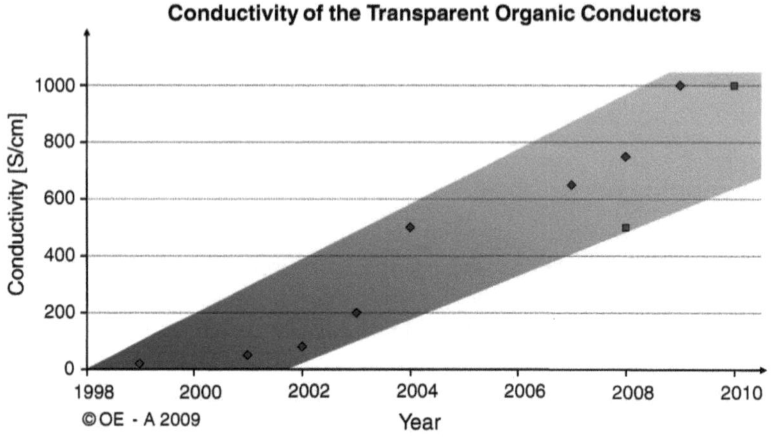

Fig. 1.13 Progress of the electrical conductivity of PEDOT:PSS-dispersions over the past 10 years. *Source* OE-A

poly-3-hexyl-thiophene (P3HT), and of the widely used molecular semiconductor pentacene [17–19]. Organic semiconductor materials are starting to be available as pre-formulated inks (see Fig. 1.15). The charge transport properties depend on both the molecular structure and the deposition conditions such as solvents, deposition technique, concentration, interfaces etc. Most of the organic semiconductors used today are p-type (like pentacene and polythiophene), but n-type materials are becoming more widespread; having both p- and n-type materials enables CMOS-type circuits, which have significant advantages, e.g. lower power consumption. The charge carrier mobility of organic semiconductors, though still much lower than crystalline silicon, has improved dramatically in recent years, already matching amorphous silicon (a-Si), and is expected to approach or match polycrystalline silicon (poly-Si) in coming years, first in research, where mobilities of up to 2.5 cm^2/Vs have already been reported, and some time later in commercial products (Fig. 1.16) [20]. This will be possible with optimized small molecule materials and polymers or new materials as e. g. inorganics, nanomaterials, carbon nanotubes or hybrid materials.

Small molecule organic semiconductors are of growing interest. These materials have usually been deposited by vacuum evaporation or other vapour-phase processes, but more recently deposition is no longer restricted to evaporation processes; several semiconductors of this type can be processed in solution or dispersion and therefore are compatible with solution coating or mass printing processes. In addition, high throughput evaporation processes might enable the large-scale use of this class of materials.

Inorganic materials such as **metal oxides** [21] or **solution processible Si** [22] have also generated much interest recently; these can be deposited by vapour phase processes or from solution as nanoparticles or precursors, with reported mobilities in the range of poly-Si for metal oxides and even higher for solution

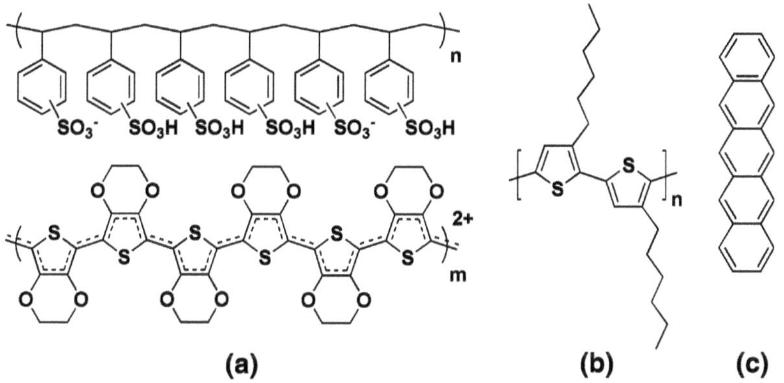

Fig. 1.14 Structures of common materials for organic electronics. **a** conductor PEDOT:PSS. **b** semiconductor polythiophene P3HT. **c** semiconductor pentacene. *Source* OE-A

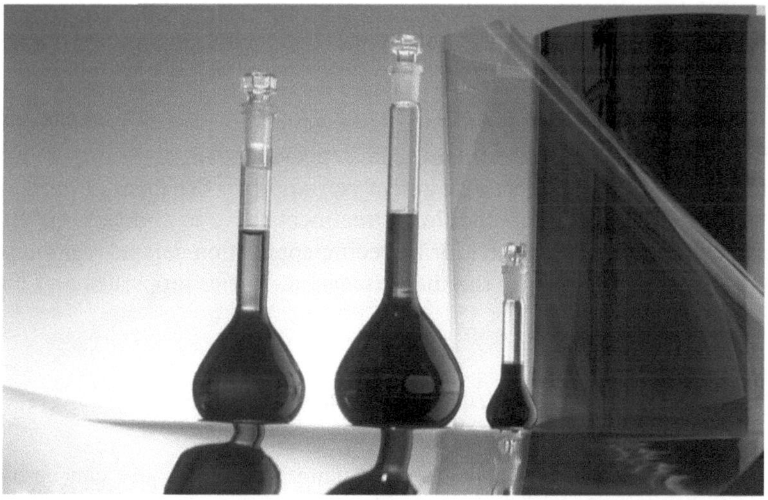

Fig. 1.15 Ready to use transparent conductive polymer solutions. *Source* H.C Starck Clevios

processible Si. An open issue with this class of materials is still the relatively high processing temperature needed to achieve high performance.

New material classes like carbon **nanotubes** or **hybrid (organic–inorganic) material** combinations could enable further improvements in the performance of the devices. Nanotubes have been used both as semiconductors and as the basis for transparent conducting films [23].

A principle advantage of organic electronics is that large, flexible and low cost **substrates** can be used. Polymer films (like the polyesters PET and PEN, or other polymers like polyimide or polycarbonate) are most widely used today, but paper, cardboard, thin glass and stainless steel are also prominent candidates. Special surface treatment or barrier layers can be added if necessary. For many

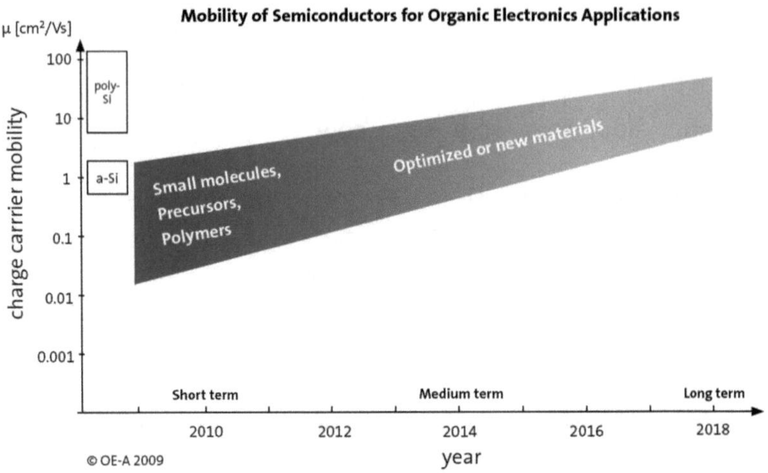

Fig. 1.16 OE-A roadmap for the charge carrier mobility of semiconductors for organic electronics applications. The values refer to materials that are available in commercial quantities; research samples may show significantly higher mobilities. The values for amorphous silicon (a-Si) and polycrystalline silicon (poly-Si) are given for comparison. *Source* OE-A

applications careful surface treatment such as planarisation, or coating with barrier materials is necessary. Pre-heat treatment can improve the thermal properties of some substrates. However, all additional treatments have of course some effect on the cost. The material best suited for a specific application depends on the process conditions, surface roughness, thermal expansion, barrier properties and cost.

1.3.2 Printing and Patterning Techniques

A wide range of large area deposition and patterning techniques can be used for organic electronics. Most prominent in this context are various printing techniques that are well known from the graphic arts industry and enable reel-to-reel processing. An in-depth analysis of these techniques and their application to organic device manufacturing is given in Chap. 2 of this book (Konarka).

Examples of two high volume printing processes, **gravure** and **screen**, are shown in Figs. 1.17 and 1.18.

Other **mass printing processes** are **offset lithography** and **flexography.** The lateral resolution (smallest feature that can be printed) typically ranges from 20 to 100 μm depending on process, throughput, substrate and ink properties, but there has been recent progress on moving to feature sizes as small as 10 μm. Film thicknesses can range from well under 1 to 10 μm. These printing processes can have enormous throughput and low production cost, but place demanding requirements on the functional inks in terms of properties like viscosity, and

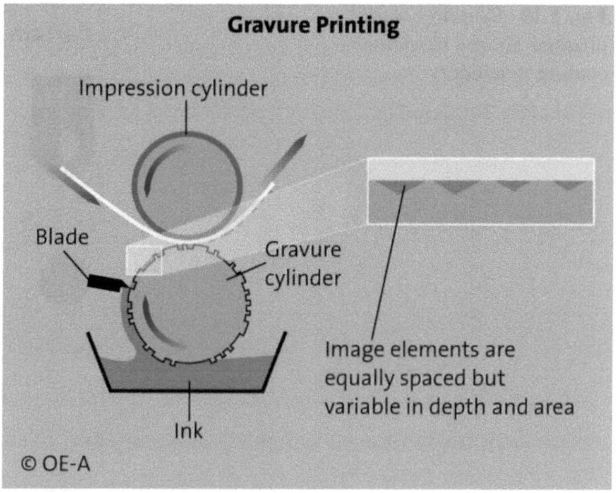

Fig. 1.17 Gravure printing process. *Source* OE-A

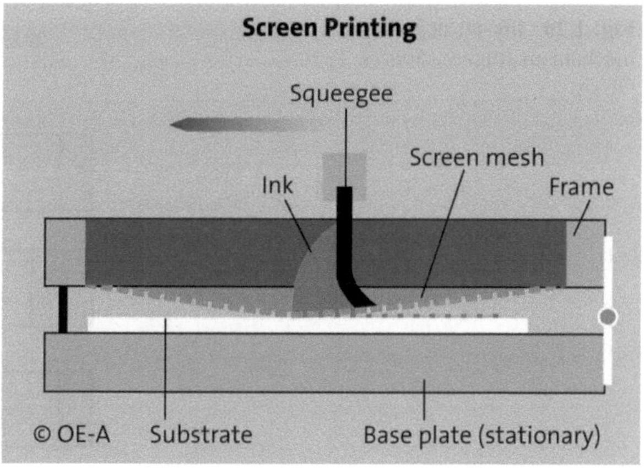

Fig. 1.18 Screen printing process. *Source* OE-A

cannot correct for issues like substrate distortion. Mass printing will be an important production process especially for applications where large area, high volumes and low costs are important. Related to volume printing are unpatterned **solution coating techniques** such as **slot-die, wire bar** or **curtain coating** (Fig. 1.19).

Ink-jet printing has received growing interest as a way to deposit functional materials (see Fig. 1.20). Being ink-jet a digital printing process, where no printing plate is needed, this technique enables variable printing and can correct in-line for distortions. Ink-jet printing head developers have continued to manufacture finer and finer printing heads, which are starting to enable features on the order of a few µm, and throughput is improving with the development of multi-head printers.

Laser ablation, large area vacuum deposition, soft lithography and large area photolithography are **further patterning and deposition techniques.** Some of these processes are subtractive, i.e. involve removing unwanted material from a

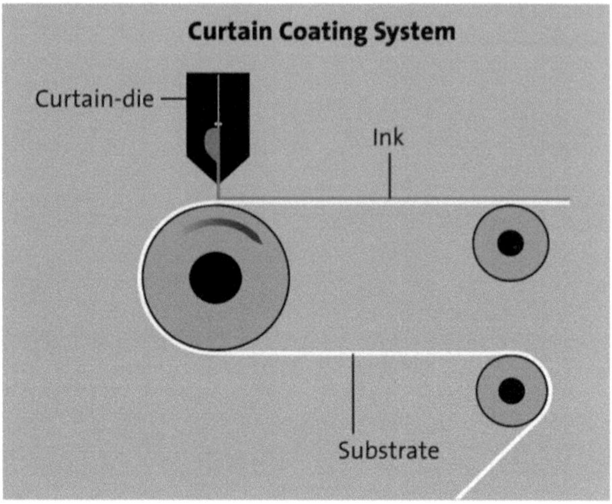

Fig. 1.19 Curtain coating process. *Source* Coatema coating machinery

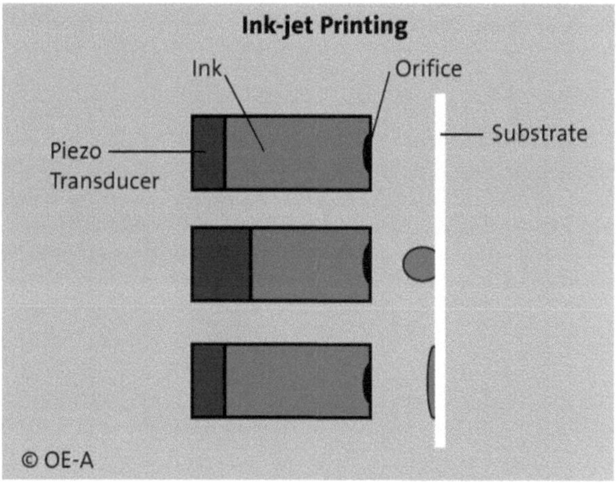

Fig. 1.20 Ink-jet deposition mechanism (piezo). *Source* OE-A

large area unpatterned film, while others are additive, i.e. only deposit material where it is wanted. Sub-μm patterning techniques such as nanoimprint lithography and micro contact printing have gained a good deal of attention recently but are still primarily used in research. Each method has its individual strengths, and in general, processes with a higher resolution have a smaller throughput (Fig. 1.21).

There are no single standard processes in existence today. Deciding which printing or other patterning process is used depends on the specific requirements of a particular device. In general, different processes have to be used for subsequent steps of a multilayer device in order to optimize each process step. The above mentioned processes differ strongly with regard to e.g. resolution and throughput, and one system may require some high throughput steps followed by high resolution processes, e.g. deposition of large amounts of material using coating or mass printing followed by fine patterning of a small portion of the surface using laser ablation.

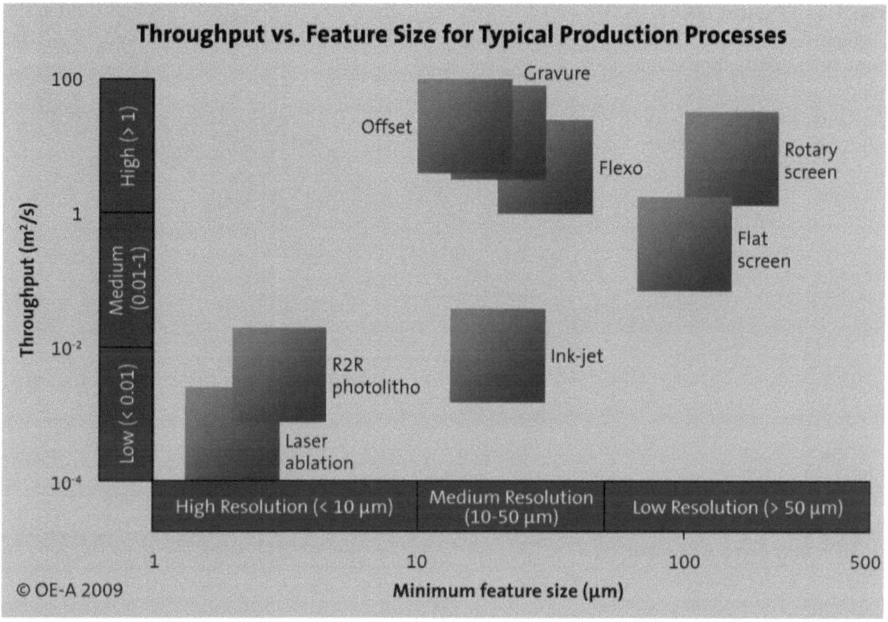

Fig. 1.21 Throughput versus feature size for a range of printing and patterning processes. *Source* OE-A

1.3.3 Devices

The organic materials can be combined to a number of **active components** such as transistors, diodes, various types of sensors, memories, photo-voltaic cells, displays or batteries. Examples for **passive devices** are conductive traces, antennas, resistors, capacitors or inductors.

Transistors are a key component of many electronic devices, including RFID or O-TFT backplanes (Organic Thin Film Transistors) for displays, and are a building block for most electrical circuits. An example of the configuration of a typical organic field-effect transistor is shown in Fig. 1.22. Essentially, the device consists of four layers: gate electrode, insulator, source/drain electrodes and the semiconductor. The current flow between source and drain electrode is switched, depending on the voltage applied at the gate electrode. In order to optimize the transistor properties, the channel length should be as small as possible and the mobility of the organic semiconductor should be as high as possible.

The other key active components in organic electronics are **diodes.** These can be large-area devices such as **OLEDs** based on small molecules or polymers (Fig. 1.23) and **photovoltaic cells**, or small area components in a circuit. In particular, rectifying diodes are a key component in RF circuits [24] and recently have been demonstrated in display backplanes [25] and memory cells [26] as well. Typically a diode consists of two electrodes (one of them transparent for

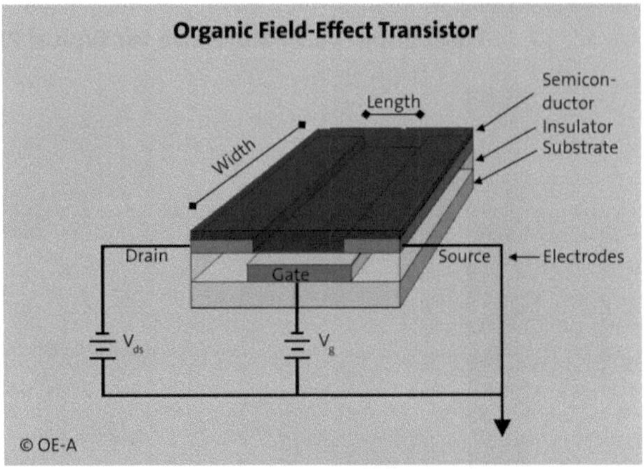

Fig. 1.22 Typical *OFET* (organic field-effect transistor) configuration and connections. The thickness of the layer stack is typically below 1 μm. *Source* OE-A

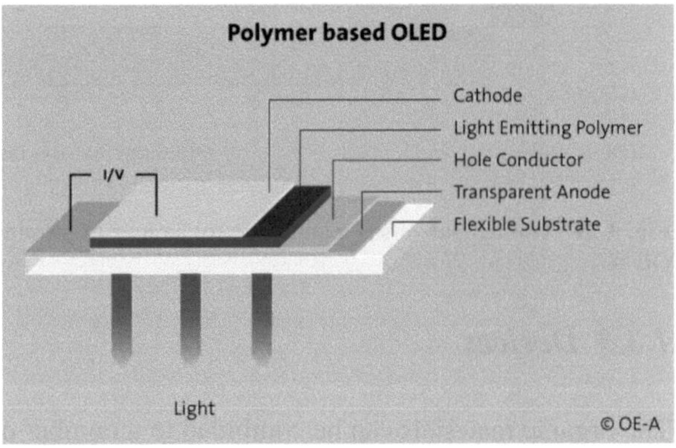

Fig. 1.23 Typical configuration for a polymer based OLED. Small molecule OLEDs may have a number of layers with different functions. The thickness of the layer stack is typically below 1 μm. *Source* OE-A

photovoltaic cells or OLEDs) and anywhere between one and several organic layers with different functions such as hole or electron transport, light absorption or light emission. More details on organic photovoltaic cells and OLEDs can be found respectively in the Chaps. 2 (Konarka), 3 (Samsung), and 4 (Philips) of this book.

1.3.4 Technology Levels

The technologies that are used in organic electronics range from batch, clean-room, etching based processes to mass printing processes that are capable of deposition of square meters of substrates per second.

Here is a rough classification of the technologies in three different technology levels:

The **wafer level** technology includes batch processing, typically using film substrates on a carrier. An adapted semiconductor line is used for processing. High resolution can be achieved by vacuum deposition and/or spin coating followed by photolithography and wet or dry etching. The production cost is relatively high and the process is not compatible for conversion to in-line sheet to sheet or reel to reel processes.

Under **hybrid** technologies, we summarize combinations of processes including large area photolithography, screen printing or printed circuit board (PCB) technologies that make use of flexible substrates (e.g. polymer films or paper). Deposition of materials is by spin coating, doctor blading or large area vacuum deposition, in some cases also partly by printing. Ink-jet printing and laser-patterning are further technologies that are grouped in the hybrids and enable production at a medium cost level.

Fully printed technology is the term we use to refer to the use of continuous, automated mass-production compatible printing and coating techniques, flexible substrates and high throughput sheet to sheet or reel-to-reel processing (see Fig. 1.24). Although all-printed devices do not yet show as high resolution or performance as those made using wafer or hybrid processes, mass printing has great potential for very low cost production and will be able to deliver extremely large numbers of products. At the same time it requires significant volumes of materials even for trials, and will need large volume applications to properly utilise such high-throughput equipment.

1.3.5 Key Technology Parameters

The detailed application parameter specifications for the different applications and product generations help define the requirements that have to be fulfilled from the technology side. The technology parameters are more "fundamental" and describe fundamental material, device or process properties. As with the application parameters, we only list a small excerpt of the key technology parameters identified for the various applications, focussing on those that are relevant to a number of applications.

- **Mobility/electrical performance**

(threshold voltage, on/off current ratio)

The performance (operating frequency, current driving capacity) of the circuits depends on parameters like the carrier mobility of the semiconductor and the threshold voltage, the conductivity of the conductor and the dielectrical behaviour of the dielectric materials.

- **Resolution/registration**

The performance (operating frequency, current driving capacity) and reliability of the circuits depends on the lateral distance of the electrodes (resolution) within the devices (e. g. transistors) and the overlay accuracy (registration) between different patterned layers.

Fig. 1.24 Reel-to-reel flexographic printing of electronic devices. *Source* Acreo

- **Barrier properties/environmental stability**

The lifetime depends on a combination of the sensitivity of the materials and devices to oxygen and moisture and the barrier properties of protective layers, substrates and sealants against oxygen and moisture. The necessary barrier properties vary for the different applications over several orders of magnitude.

- **Flexibility/bending radius**

Thin form factors and flexibility of the devices are key advantages of organic electronics. In order to achieve reliable flexible and even roll-able devices materials, design and process have to be chosen carefully.

- **Compatibility of process parameters**

(speed, temperature, solvents, ambient conditions, vacuum, inert gas atmosphere)

In order for a multi-component system to work properly and be easily manufacturable, it is important to adjust the parameters of the different materials and devices and choose the right order of processing.

- **Yield**

Low cost electronics in high volumes are only possible when the processes allow production at high yields. This requires reliable and validated processes, optimised materials and circuit designs as well as an in-line quality control.

1.4 Main Challenges

One goal of the roadmap is to identify **red brick walls**—principle challenges that can only be overcome by major breakthroughs beyond the expectations of standard technology development. For each application the requirements for product generations were compared with expected technology development and the key challenges were identified and discussed. Like the key application and technology parameters, the red brick walls may vary for the different applications. Those discussed below are the most important ones and are relevant for all applications.

A common feature of all future generations of the different products is that the **complexity and overall size of analogue and digital circuits** is increasing. In certain cases, the applications include millions of transistors, other combine various different electronic devices like transistors, power supply, sensors, displays and switches. In the future more and more higher and higher performance components will have to be fit into smaller and smaller areas, while for other applications high performance components will have to be placed precisely over large areas, up to a few square meters. At the same time, wafer level processing will not in the long term be a commercially viable approach for a number of applications and hybrid or fully printed processes will need to be used.

Based on the above considerations and the results of the work of the application and technology groups, we believe that **major breakthroughs** in the following areas are absolutely necessary:

resolution, registration and **process stability** of the patterning processes

charge carrier mobility and electrical conductivity of the semiconductor and conducting materials

circuit design and realization of increasingly complex circuits including complementary (analogous to CMOS in Si technology) transistors

These challenges cannot be treated in separate ways since they depend on each other. Resolution and registration accuracy differ for the various patterning techniques and even within a technique largely depend on the throughput or printing speed. The process stability depends on tolerable deviation, the circuit design and the materials that are used.

In order to enable mass production of complex devices, **resolution** better than 10 μm with as good or better registration accuracy, even on plastic substrates, is necessary. Scaling to smaller structures will become important for improved performance and increased yield, as well as to reduce the footprint for the most complex circuits. This cannot be achieved with the current level of development in high throughput, large area processes. At the same time, new strategies for **quality control** enabling high speed in-line measurement and electrical testing have to be developed. These developments will be essential to enabling low cost production at high volumes and yield.

Charge carrier mobility over 1 cm^2/Vs for processable semiconductors will be needed. These values have to be achieved in the final device using high volume processes. Charge carrier mobility in the order of 5–10 cm^2/Vs in printable commercially available materials would represent a breakthrough since it would enable more complex devices. While such values are starting to be reported on a lab scale, commercial availability could lead to breakthroughs in the industry. Further optimisation of existing materials or development of novel classes of materials will be needed to achieve this. In addition to polymers, potential candidates include small molecule and inorganic semiconductor materials as well as nanomaterials and new hybrid systems that can be processed from solution.

Another principle challenge is the **circuit design** for complex circuits that are compatible with a broad range of materials and mass printing processes. In particular, complementary circuits need to be developed, which requires high quality p- and n-type semiconductors. This will enable complex circuit designs and will significantly increase functionality of the devices, as it did previously in silicon technology. CMOS analogue design also brings significant advantages in yield, speed and available functions. In addition, designs for lower supply voltages and higher frequencies are of great importance. Progress has been made in this area recently, with gravure printed complementary inverters reported [27] and large European projects aimed at more complex applications of organic complementary circuitry [28], but much still remains to be done.

A key reason to identify red brick walls is to help the organic electronics community align its efforts to solve the most pressing problems. Long-term strategies, funding and new partnerships along the value chain are necessary to overcome the red brick walls.

1.5 Summary and Outlook

Organic and printed electronics is a new and fascinating platform technology that enables fresh electronic applications in many fields, such as interactive toys, RFID-tags, sensors, rollable displays or flexible solar cells, which are now entering the market. With this third version of the OE-A roadmap, we have updated and expanded the information about our view of the developments in this field. We included new applications, updated the status, key parameters and expected technology development for them and used this updated information to identify the key challenges, which we call red brick walls. We have also tried to bring the basic information together to a relatively simple picture of the main developments in this field from the application and technology point of view.

We have found that the technology is mature enough to enter the market with first, relatively simple products, addressing interesting market segments, and since the last edition of the roadmap new products have started to appear. We also have seen that mass markets could be reached in the near future but this will depend on progress in the fields of materials, equipment, processes, devices and circuit design. Some of this progress will be straightforward, while we have identified some areas where breakthroughs will be needed.

For example, the development of an organic CMOS-like technology could result in a breakthrough of organic electronics, just like it did with silicon electronics. Improved patterning processes and materials with better electrical performance and processability are also key for future product generations. It is expected that new organic and inorganic materials will play an important role too.

Very important are also the new developments in inline quality control of electrical parameters, especially in printing processes. This will allow sufficiently high yields to reach low cost, high volume products. Standardization in materials, processes and device design gain more and more importance as organic electronics is entering the production phase.

However, some questions remain open; for example we have not yet been able to define a simple "Moore's law" for organic electronics. Organic electronics is still a very young field, and there are still many different parameters that are important for its further success; it is not clear which of these parameters might have the most important role or how they will scale. However, there are indications that such parameters as charge carrier mobility, feature size and circuit complexity could become candidates for simple scaling laws in the future. We also have not yet been able to identify the "killer application" for organic electronics in the long term; there are many fascinating applications and time will tell which of these—or new ones we have not yet thought of—will turn into a "killer application".

Organic and printed electronics is now in the market and has great potential for further growth. We will continue to follow the developments to find the major trends. The organic electronics roadmap is an ongoing task and key activity of the OE-A and its members, and we will regularly update the roadmap and keep the community informed.

References

1. Clemens W, Lupo D, Hecker K, Breitung S (2009) White Paper OE-A Roadmap for Organic and Printed Electronics. Organic Electronics Association (OE-A). http://www.vdma.org/wps/myportal/Home/en/Datenbanken/Downloads?WCM_GLOBAL_CONTEXT=/vdma/Home/en/Datenbanken/Downloads&initsearch=&branche=OEA. Accessed 2011(Since Dec 2011 an updated (4th) edition of the OE-A Roadmap is available at this site.)
2. Menippos GmbH. http://hurrafussball.com/. Accessed 2011
3. Varta Microbatteries GmbH. (http://www.varta-microbatteries.com/en/newsandpr/news_-new/news_new.php?action = detail&id = 115). Accessed 2011
4. Konarka Technologies, Inc. www.konarka.com. Accessed 2011
5. G24 Innovations. www.g24i.com. Accessed 2011
6. Koninklijke Philips Electronics N.V. http://www.lighting.philips.com/main/lightcommunity/trends/oled/. Accessed 2011
7. OSRAM GmbH. http://www.osram.com/osram_com/LED/OLED_Lighting/index.html. Accessed 2011
8. Solaronix SA. http://www.solaronix.com/products/dyesolarcells/. Accessed 2011

9. Clemens W, Krumm J, Blache R (2010) Printed RFID and Smart Objects for new high volume applications, Ibidem, Section 6.5
10. Myny K, Beenhakkers MJ, van Aerle NAJM, Gelinck G H, Genoe J, Dehaene W, Heremans P (2009) A 128b Organic RFID Transponder chip, including manchester encoding and ALOHA anti-collision protocol, operating with a data rate of 1529b/s. ISSCC Dig Tech Papers 206–207
11. Genoe J (2008) High frequency rectification for organic RFID tag. Organic Semiconductor Conference 2008, Frankfurt Messe, Frankfurt Germany, 30 Sept 2008
12. Sutija D (2010) Commercialization path for non-volatile rewritable memories—current markets and the pathway to integrated products, LOPE-C 2010, Frankfurt Messe, Frankfurt Germany, 31 May 2010
13. Mildner W (2009) Roadmap for organic and printed electronics. LOPE-C 2009, Frankfurt Messe, Frankfurt Germany, 24 June 2009
14. Interactive Wear AG. http://interactive-wear.de/cms/front_content.php?client=1&changelang=3&parent=&subid=&idcat=72&idart=112. Accessed 2011
15. Brochier Technologies. http://www.brochiertechnologies.com/gb/index.html. Accessed 2011
16. Organic Electronics (2006) 1st edn VDMA Verlag GmbH, Frankfurt am Main, Germany
17. Wikipedia. http://en.wikipedia.org/wiki/PEDOT:PSS. Accessed 2011
18. Wikipedia. http://en.wikipedia.org/wiki/P3HT. Accessed 2011
19. Wikipedia. http://en.wikipedia.org/wiki/Pentacene. Accessed 2011
20. Llorente GR, Dufourg-Madec M-B, Crouch DJ, Pritchard RG, Ogier S, Yeates SG (2009) High performance, acene-based organic thin film transistors. Chem Commun 21:3059
21. Park S-J, Jeong JK, Mo Y-G, Kim S (2009) Impact of high-k TiOx dielectric on device performance of indium-gallium-zinc oxide transistors. Appl Phys Lett 94:042105
22. KOVIO. http://www.kovio.com/index.html. Accessed 2011
23. Arias AC, Mackenzie JD, Rivnay J, Salleo A (2010) Materials and applications for large area electronics: solution-based approaches. Chem Rev 110:3–24
24. Myny K, Steudel S, Vicca P, Genoe J, Heremans P (2008) An integrated double half-wave organic Schottky diode rectifier on foil operating at 13.56 MHz. Appl Phys Lett 93:093305
25. Lilja KE, Bäcklund TG, Lupo D, Virtanen J, Hämäläinen E, Joutsenoja T (2010) Printed organic diode backplane for matrix addressing an electrophoretic display. Thin Solid Films 518(15):4385–4389
26. Naber RCG, Asadi K, Blom PWM, de Leeuw DM, de Boer B (2010) Organic nonvolatile memory devices based on ferroelectricity. Adv Mater 22:933–945
27. Yan H, Chen Z, Zheng Y, Newman C, Quinn JR, Dötz F, Kastler M, Facchetti A (2009) A high-mobility electron-transporting polymer for printed transistors. Nature 457:679–686
28. POLARIC. http://www.vtt.fi/sites/polaric/ and COSMIC -http://www.project-cosmic.eu/project.html. Accessed 2011

Chapter 2
Solution-Processed Organic Photovoltaics

Claudia N. Hoth, Pavel Schilinsky, Stelios A. Choulis, Srinivasan Balasubramanian and Christoph J. Brabec

Abstract The technology of organic solar cells has matured to an extent that commercialization of first products has already started. However, with the first products pushing into the market, the research community realizes that a qualified product requires more than only high efficiency and good stability. Cost is of course as important as efficiency and lifetime, but to achieve high productivity, multiple technologic challenges have still to be solved. To reduce production costs, printing of functional layers from solution has evolved to a promising manufacturing technology for flexible organic electronics. Current processing of organic photovoltaic devices is mainly based on traditional methods like spin coating or doctor blading. However, these techniques have several disadvantages such as the incompatibility with a roll-to-roll setup and the processing of only small areas at laboratory scale. Enormous benefits in the manufacturing of organic photovoltaics are achieved by using low-cost roll-to-roll capable technologies including screen printing, spray coating, inkjet printing, gravure/flexographic printing and curtain/slot die coating. This review will shed some light on the role and importance of production technologies for organic photovoltaics and give an update on the most recent achievements in the field.

C. N. Hoth (✉) · P. Schilinsky
Konarka Technologies GmbH, Landgrabenstrasse 94, 90443 Nürnberg, Germany
e-mail: choth@konarka.com

S. A. Choulis
Department of Mechanical Engineering and Materials Science and Engineering, Cyprus University of Technology, 3603 Limassol, Cyprus

S. Balasubramanian
Konarka Technologies Inc., 116 John Street, Lowell, MA 01852, USA

C. J. Brabec
Institute Materials for Electronics and Energy Technology (I-MEET), Friedrich-Alexander-University Erlangen-Nürnberg, Martensstr. 7, 91058 Erlangen, Germany

E. Cantatore (ed.), *Applications of Organic and Printed Electronics*,
Integrated Circuits and Systems, DOI: 10.1007/978-1-4614-3160-2_2,

Keywords Organic solar cells · OPV · Printing · Coating · Polymer · Fullerene · Bulk heterojunction · Device fabrication · Lifetime

2.1 Introduction

Amorphous semiconducting polymeric materials have attracted significant interest in the field of organic electronics due to their processing advantages [1–3]. The possibility of applying these materials at low temperatures using solution coating techniques has resulted in intense research activity in recent years. The developments in the fields of organic field effect transistors (OFET or OTFT), organic light emitting diodes (OLED), organic photodetectors and organic photovoltaics (OPV) have now made organic electronics commercially viable. The low-temperature solution processability allows for roll-to-roll printing or coating on flexible plastic substrates which is expected to lead to light-weight, low-cost electronic devices including among others displays, photovoltaic devices and thin film batteries.

For photovoltaics in particular, roll-to-roll production is attractive due to the requirements of low cost, light weight and large area coverage that characterize these products.

Other technologies such as thin film inorganic semiconductor devices that require, for instance, chemical vapour deposition or lithography do not realize the vision of low-cost products due to higher manufacturing cost when compared to printing and coating.

Inorganic bulk semiconductors do not, or only in a very limited way, offer a processing window adequate for solution processing. Inorganic solar cells based on colloidal semiconductor nanocrystals (CdTe and CdSe) spin coated from solution have been presented by the Alivisatos group. These air-stable devices performed well with 2.9 % power conversion efficiencies (PCE) [4]. Inorganic solution-fabricated CIGS ($Cu(In,Ga)Se_2$) cells are being commercialized with over 10 % PCE [5, 6], but these materials suffer from complex processing, e.g. high temperature conditions, which do not enable the use of low-cost flexible plastic substrates and roll-to-roll processing. Roll-to-roll vacuum processing of thin film inorganic photovoltaic based on materials like amorphous silicon has also been demonstrated. However, this approach is characterized by much lower throughput and significant higher costs than printing technologies.

In addition to production costs, the competitive position of a photovoltaic technology must be assessed comparing how much efficiency can be gained per spent dollar with the different approaches.

In this chapter we analyze the potential of low-cost printed organic photovoltaics. The challenges associated with large-scale printing or coating procedures are addressed together with an overview of OPV performance in terms of efficiency and operational stability.

At present bulk heterojunction (BHJ) composites based on blends of a polymer donor and a fullerene acceptor for ultrafast charge transfer at the donor/acceptor

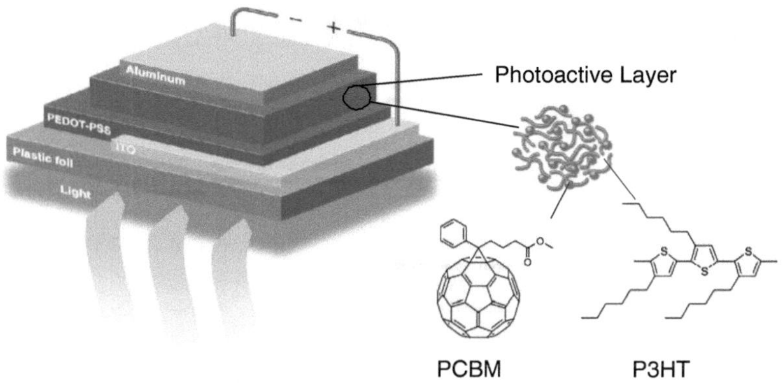

Fig. 2.1 Organic solar cell device structure. The photoactive layer consists of a blend based on poly(3-hexylthiophene) (P3HT) and the C60 derivative PCBM ([6,6]-phenyl-C61-butyric acid methyl ester)

interface are the organic material system with the highest reported efficiencies. The working horses for electron donating and accepting materials in the BHJ structure are the well-known poly(3-hexylthiophene) (P3HT) and the C60 derivative PCBM ([6,6]-phenyl-C61-butyric acid methyl ester). The chemical structures of the materials as well as a typical device configuration are shown in Fig. 2.1.

For a solar cell at least one transparent electrode is required, which is typically a conductive oxide (TCO). The transparent electrode is often defined by indium tin oxide (ITO) coated on glass or flexible plastic carriers. Together with a thin layer of the intrinsically conductive poly(3,4-ethylene dioxythiophene) doped with the polyanion polystyrene sulfonate (PEDOT:PSS), a high work function hole-collecting electrode is built. The photoactive layer is formed by the donor–acceptor blend film, which forms morphologies with phase separation in the nm-scale characterized by good percolation pathways for efficient charge collection and decreased charge recombination. The use of two materials emphasizing different electronic levels is one of the most important design concepts for organic bulk heterojunction solar cells. With this donor–acceptor material combination several groups have reported laboratory device efficiencies in the range of 4 % [7–9]. Low-bandgap benzothiadiazole-fused thiophene copolymers such as poly[2,6-(4,4-bis-(2-ethylhexyl)-4*H*-cyclopenta [2,1-b;3,4-b']-dithiophene)-*alt*-4,7-(2,1,3-benzothiadiazole)] (PCPDTBT) used as donor materials represent a promising route to improved device efficiency, due to the better overlap between the PCPDTBT absorption spectrum and the solar emission spectrum [10–14]. A new class of polymer materials which are viable for OPV manufacturing currently give certified performance beyond 8 % [15]. In this chapter, however, we focus our analysis on the well known and widely used P3HT:PCBM blends.

The origin of the open circuit voltage (V_{OC}) is a matter of controversial discussions. V_{OC} is the voltage at the terminals of a solar cell when no current flows and is basically determined by the distance between the highest occupied molecular orbital

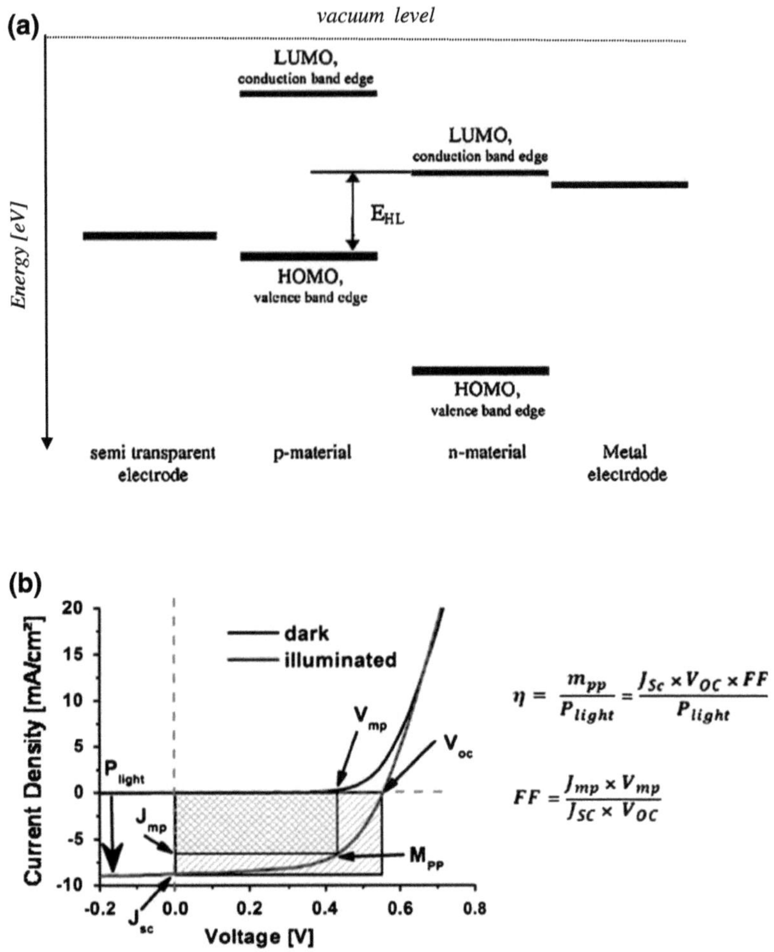

Fig. 2.2 **a** Energy band scheme of a bulk heterojunction solar cell. The V_{OC} is determined by the distance of the highest occupied molecular orbital (*HOMO*) level of the donor and the lowest unoccupied molecular orbital (*LUMO*) level of the acceptor (indicated as E_{HL}). **b** Characteristic *J–V*-curve of a solar cell in the dark (*upper curve*) and under illumination (*lower curve*). The parameters J_{SC}, V_{OC}, M_{PP} and the power of the incident light P_{light} are depicted. The ideal power of a solar cell is represented by the *dashed square*. The real maximum power is shown by the *checkered square*. Moreover, the relationship between device performance, V_{OC}, J_{SC} and *FF* is given

(HOMO) level of the donor and the lowest unoccupied molecular orbital (LUMO) level of the acceptor material, as shown in Fig. 2.2a (indicated as E_{HL}).

Under illumination a photo-induced charge transfer occurs. As the bulk absorbs photon energy, coulombically bound electron–hole-pairs (excitons) are generated, that dissociate at the donor–acceptor interface into electrons and holes. Charge transport to the opposite electrodes takes then place: holes are transported by the p-type material to the higher work function electrode (TCO) and electrons by the n-type material to the lower work function contact (metal), respectively. The short circuit current density (J_{SC}) depends on the number of photogenerated charges collected by the electrodes. The J_{SC} is among others affected by the absorption of photons, the mobility of charge carriers within the active layer, recombination processes and the morphology of the active layer. The relationship between device efficiency, V_{OC}, J_{SC} and fill factor (*FF*) is shown in Fig. 2.2b on a typical *J–V* characteristics under illumination and in the dark. The *FF* is defined as the ratio of the actual maximum obtainable power, ($V_{mp} \times J_{mp}$) to the maximum theoretical power, ($J_{sc} \times V_{oc}$). The *FF* is limited by insufficient transport properties, increased series resistance R_s or reduced parallel resistance R_p. R_s and R_p can be determined by the inverse of the slope of the *J–V* characteristics in the points V_{oc} and J_{sc} respectively.

The performance of BHJ solar cells depends critically on the material properties and processing conditions. Depending on the printing technology the ink must fulfil certain requirements for a reliable printing and adequate film formation in terms of viscosity, boiling point, vapour pressure, polarity and surface tension. The selection of organic solvents is determined by solubility, drying characteristics, fluidity, surface tension and environmental and regulatory considerations. Therefore, the ink is formulated with respect not only to the solvent properties, but also to adapt the ink to the printing technology. The characteristic properties of the different printing methods applied to manufacture solar cells and the requirements on the inks will be extensively discussed in the Sect. 2.2.

Section 2.3 will concentrate on reliability and yield aspects of OPVs, while Sect. 2.4 will discuss accelerated lifetime test procedures and Sect. 2.5 will present a possible roadmap for printed solar cells.

2.2 Production Technologies of Organic Photovoltaics

In recent years significant progress has been made in the processing of organic bulk hetero-junction solar cells. In this section we discuss the different solar cell fabrication technologies such as doctor blading, screen printing, spray coating, inkjet printing, gravure and flexographic printing, curtain and slot-die coating as compared to the conventional spin-coating process. These various technologies can all enable roll-to-roll printing of organic solar cells. The high throughput and productivity (amount of generated energy per energy used during manufacturing) of roll-to-roll (R2R) printing is expected to enable low-cost commercial solar cells. Due to the high level of development and the variety of printing and coating methods available, high resolutions down to 20 microns as well as high layer qualities for a broad range of printable materials can be achieved [16]. A roll-to-roll process simplifies manufacturing scale-up and has significantly lower costs than other methods to make solar cells. Printing processes are non-toxic, environmentally friendly and, being compatible with low temperatures, are less energy intensive than 1st or 2nd generation photovoltaic technologies [17]. Another significant advantage of roll-to-roll processes is that they can be implemented using existing coating and printing equipment, and thus does not require the construction of a new infrastructure [17].

Before discussing the details of the individual printing methods, it is essential to recall that the quality of any printing method heavily depends on the degree of optimization of the ink formulation. Some of the key requirements for organic solar cells including uniform film quality with low density of defects (pinholes or line defects) over large areas necessitate ink formulations with appropriate rheological properties. The ink should provide a long shelf life and high consistency in concentration, temperature behaviour and viscosity. Moreover, the substrate should exhibit high uniformity and reliable surface properties and during the printing process the web speed and the tension at the point of ink deposition has to be uniform.

Another important aspect of ink formulation is the drying kinetics, which is actually quite independent of the specific printing technology used. In general, one is interested to work with fast drying inks, which allow high web speeds and shorter drying stages. For OPV production, requiring solid films in the range of tens or few hundreds of nanometers and a typical wet film thickness in the 10 micron regime, inks with high drying rates are preferred. Conversely, the use of high-boiling-point organic solvents, which can delay the drying process significantly, should be avoided.

2.2.1 Spin Coating

Early reported organic photovoltaic devices were fabricated by spin coating [18, 19]. Spin coating is a well-understood laboratory method to deposit organic semiconductor solutions onto substrates resulting in high quality thin films. The fluid is deposited at the centre of the substrates. Spinning of the substrate at high velocity (up to 10,000 rpm) causes centrifugal acceleration and thus, spreading of the fluid on the surface. The final film thickness depends not only on the nature of the fluid in terms of viscosity, drying rate, solid concentration and surface tension, but strongly on the spinning parameters such as spinning speed, acceleration and time interval. Most of the recent research efforts in the OPV field have focused on spin coating due to the film homogeneity and low defect density that this technique enables. Spin coated cells with high power conversion efficiencies over 6 % have been reported [9, 20–23]. Unfortunately though, this method is not scalable to large volume production. Each substrate has to be handled individually and patterning the layers would require costly subtractive patterning steps like lithography. In addition most of the applied solution is wasted and only a small amount of it is eventually used to cover the substrate.

2.2.2 Doctor Blading

A couple of years after the first published spin coated organic photovoltaics, doctor blading was used to obtain thin films from solution [24, 25]. A schematic representation is shown in Fig. 2.3a.

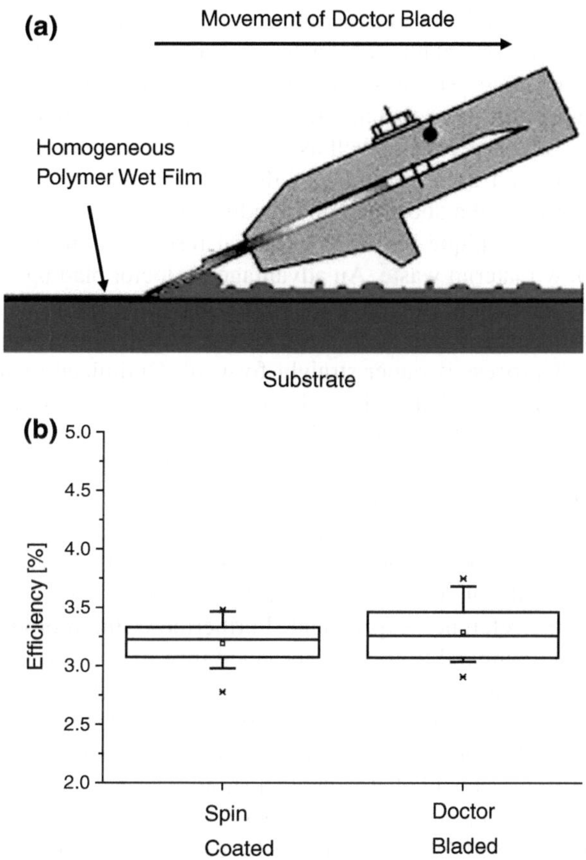

Fig. 2.3 **a** Principle of the film formation for films deposited by doctor blading. **b** Power conversion efficiencies of a series of organic solar cell devices by the state-of-the-art technique spin coating compared to the doctor blading method [26]. Copyright Wiley–VCH Verlag GmbH & Co. KGaA. Reproduced with permission. Data are presented in box plots. The *horizontal lines* in the box denote the 25th, 50th, and 75th percentile values. The error bars denote the 5th and 95th percentile values. The *two symbols below* and *above* the 5th/95th percentile error bar denote the highest and the lowest observed values, respectively. The *open square* inside the box denotes the mean value. The height of the box is the measure for the tolerance. Devices were illuminated with 100 mW cm^{-2} by a solar simulator. Efficiencies were corrected for the mismatch factor of the illumination source and represent AM1.5G values

Doctor blading basically uses a coating knife to apply a solution evenly over the substrate. A defined amount of the solution is applied either directly onto the substrate or, in case of low viscosities, in the gap between the substrate and the blade. Afterwards the knife is moved along the substrate with an adjustable velocity up to 70 mm s^{-1}. A wet thin film is left behind, which dries due to the evaporation of the solvent. Compared to spin coating, the loss of coating solution is strongly reduced. The different drying kinetics of the components may lead to distinct solid-state topographies in blend film morphologies [26]. The film formation can also be influenced by a temperature control of the substrate obtained with a table heater. The blade can be located as low as 20 microns above the surface of the substrate. The target is to achieve a layer with a designated film thickness. In almost the same manner to spin coating, the designated film thickness is a result of the ink properties as well as the blading parameters such as slit height between substrate and knife, volume of the solution that is applied in the slit, temperature control of the substrate, and blading speed. This blading method is a large-area coating technique resulting in lateral resolutions in the sub-millimeter regime and in low material waste. An advantage of doctor blading is among others the large choice of organic solvents, thus little ink formulation work is required. While spin coating is incompatible with R2R processing, the transfer of doctor blading to a R2R process is rather straight forward. Optimized organic solar cells manufactured by doctor blading have been reported with performances comparable to those prepared by spin coating [26, 27] A comparison of the power conversion efficiency (PCE) achieved by doctor bladed solar cells and spin coated ones, all prepared on glass substrates in our laboratories, is shown in Fig. 2.3b [26]. Three series, each with 16 solar cells, were compared to evaluate the quality of the blading process. As demonstrated in the box plots, the average performance of devices made with blading is slightly higher than the spin coated solar cells fabricated with the reference experiment. Furthermore, the reproducibility of the doctor blading is comparable to the reproducibility of the spin coating process. However, doctor bladed films usually show a small gradient in film thickness, which leads to different short circuit currents for cells fabricated from different areas of the film. Thus, along with a 10 % higher efficiency as compared to spin coated devices, the doctor bladed cells also show a larger variation in J_{SC}.

Similar quantum efficiencies for charge generation and charge separation in spin coated and doctor bladed OPVs are shown by external quantum efficiency (EQE)[1] measurements. An analysis of the current density–voltage (J–V) behaviour of bulk heterojunction solar cells was then carried out to examine the origin of the slight increase in device performance observed in doctor bladed devices compared to spin coated OPVs.

An equivalent circuit diagram representing the macroscopic model for describing a solar cell with one diode is presented in Fig. 2.4. The total current[2] I at a given voltage V is the algebraic sum of three contributions: the current through the diode, represented by the parameters n and J_0, the current due to the shunt resistance R_P, and the photogenerated current $I_{photogenerated}$. The current is also influenced by the series resistance in the bulk of the semiconductor and in the connections, R_S. The standard replacement circuit shown in Fig. 2.4 leads to the I–V characteristics following the relations of the equation given in Fig. 2.4 as well. Herein, I represents the measured current throughout the whole device, J_0 stands

[1] External quantum efficiency (EQE) is the percentage of the number of electrons extracted out of a solar cell per incident photon.

[2] In this chapter, following the widespread convention, J is the symbol of a current density, while I stands for a current.

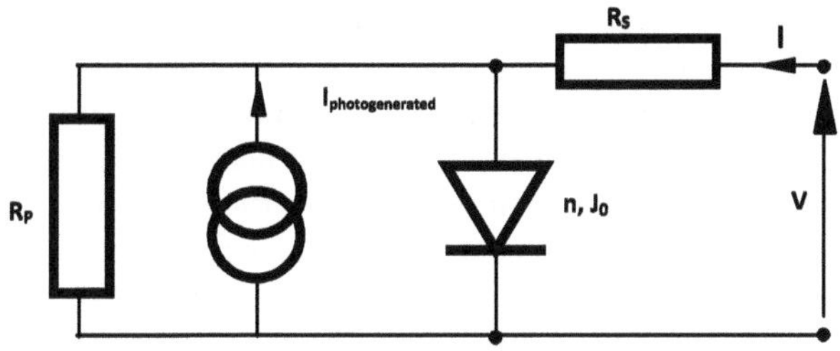

$$I = -I_{photogenerated} + AJ_0\left(e^{\frac{q(V-IR_s)}{nKT}} - 1\right) + \frac{V - IR_S}{R_P}$$

Fig. 2.4 Equivalent circuit diagram representing a macroscopic model for describing a solar cell with one diode. The total current at a given voltage V is the sum of different contributions: the current through the diode, which is represented by the parameters n and J_0, the current due to the shunt resistance R_P, and the photogenerated current $J_{photogenerated}$. The series resistance limits the total current

for the saturation current density of the diode, and n is the ideality factor of the diode. V is the applied voltage, $R_{S,\ P}$ the serial and parallel resistance, $I_{Photogenerated}$ the photogenerated current of the device, and A is the cross-section of the device.

According to the standard replacement circuit, the slope of the dark I–V curve in the voltage range between 0.2 and 0.7 V represents the diode behaviour as governed by the diode ideality factor n and the saturation current J_0 [28, 29]. The ideality factor n correlates with the number of distributed donor–acceptor interfaces within the blend [30], which is directly related to the morphology of the blend. Variations in the absolute values of n are related to different morphologies: $n = 1$ is typical of a bilayer system, while $n = 2$ represents an intimate mixing of the donor with the acceptor throughout the bulk. Values larger than two indicate more complex carrier recombination mechanisms [30]. The spin coated and doctor bladed devices based on pristine o-xylene solvent investigated in [26] and Fig. 2.3 show similar n, but different J_0, which may indicate different quality of the interfaces in the different processing techniques [26]. The reduction in J_0 in bladed devices determines a slight increase in open circuit voltage (by 30–40 mV) and thus a slight PCE enhancement [26].

The distinct drying kinetics of the wet film processed by spin coating or doctor blading leads to different solid-state morphologies, without losing performance. In conclusion, doctor blading is a better technology approach to the manufacturing of OPVs than spin coating, considering that the former is compatible with roll-to-roll processing and results in comparable or slightly increased performance [26].

2.2.3 Screen Printing

Shaheen et al. [31] presented screen printing as a fabrication technique for organic photovoltaics. A screen is patterned with an image by closing selected areas. For the deposition of the ink, the screen is located a couple of millimetres above the surface of the substrate and the ink is loaded onto the screen. Upon imprint, the squeegee transfers the ink through the open meshes onto the substrate by moving the squeegee with a velocity of several centimeters per second across the screen, momentarily touching the substrate. A wet film with designated structure/image is left behind that dries by the vaporization of the solvent formulation. Screen printing usually requires 10–100 times higher ink viscosities compared to spin coating or doctor blading. Screen printing inks show thixotropical behaviour: they are highly viscous in the resting state and become less viscous with applying a shear stress, rapidly recovering viscosity when the shear stress discontinues. To achieve such fluid behaviour, the commercial available screen printing pastes comprise among others thixotropic and viscosity modifiers, binders and thickeners to allow a reproducible and reliable screen printing process. Up to now there has been no report about the impact of these additives on the solar cell device performance. The major limitation of screen printing is defined by the ink requirements in terms of the high viscosity and stability of the ink formulation. Either a high molecular weight material with high solubility forming a stable solution is needed (which results in thicker films) or additives must be used to increase the viscosity of the ink. Nevertheless, the first organic solar cells on glass substrates with a screen printed photoactive layer were reported already in 2001 with a device performance in the range of 4.3 % PCE measured under monochromatic illumination [31]. Other groups presented screen printing as a deposition method for conjugated polymer based materials [32–34] and obtained a maximum power conversion efficiency of 1.25 % under AM1.5G spectra at 100 mW/cm^2 intensity [34].[3] Recently, screen printing of anode and cathode materials with high resolution has also been reported [35, 36].

All these results prove the potential of screen printing for manufacturing OPVs. Moreover, screen printing is an elegant and fast technology that is compatible with roll-to-roll processing.

2.2.4 Spray Coating

Spray coating [37, 38] was recently reported as a convenient technique for the fabrication of BHJ devices. Vak et al. [39] showed that spray coating the active layer from chlorobenzene solutions gives efficient devices with 2.83 % PCE. Then, Green et al. presented spray deposited P3HT:PC$_{61}$BM films from a variety of common organic solvents with different boiling points [40]. Later, Steirer et al. reported on spray coated P3HT:PCBM OPVs using chlorobenzene instead of

---

[3] Solar cells are typically measured at AM1.5G and a light intensity of 100 mW/cm^2. The "Air Mass" (AM) factor is defined as the quotient between the actual optical path length of sunlight and the optical path length when the sun is directly overhead; AM1.5G characterizes white light with a spectral intensity distribution matching that of the sun rising at a tilt angle of 37° on the earth's surface.

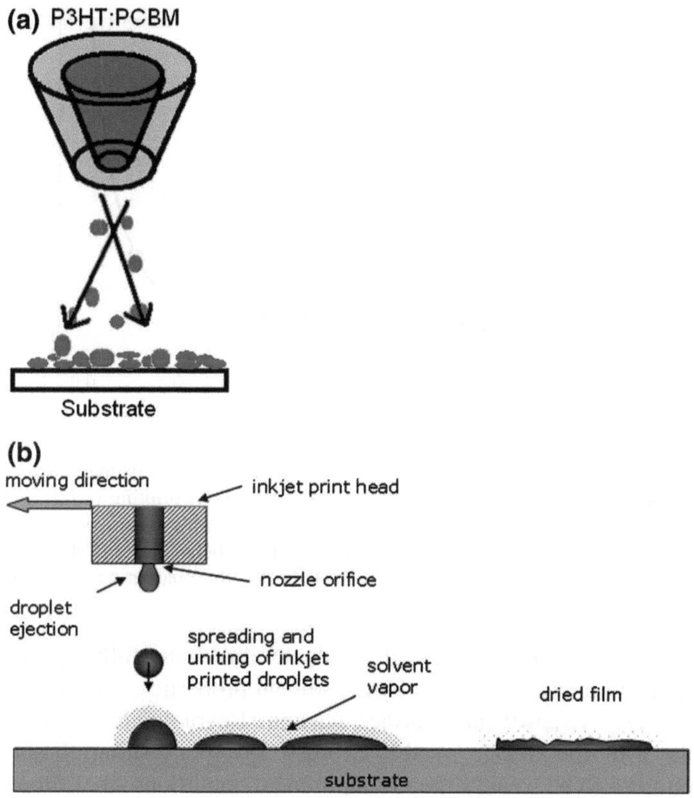

Fig. 2.5 Principle of the film formation for films deposited by (**a**) spray coating [43] and (**b**) inkjet printing [44]. Copyright Wiley–VCH Verlag GmbH & Co. KGaA. Reproduced with permission

p-xylene as solvent and found distinct morphologies for these two solvent formulations [41]. Recently, Girotto et al. investigated the influence of the airbrush settings on the surface topography of P3HT:PCBM films [42].

The film formation by spray coating is illustrated schematically in Fig. 2.5a [43]. In the spray coating technique organic thin films are generated stepwise. Single droplets are deposited by the transfer gas pressure with a high velocity onto the substrate. The droplets dry immediately when hitting the substrate surface. This results in a different morphology compared to more conventional technologies such as inkjet printing (Fig. 2.5b) [44], where the film formation is based on the spreading of droplets and combining with adjacent droplets, forming a liquid bulk that dries during vaporization of the solvent vehicle. The transfer of rather large droplets from microns to tens of microns via spray involves an individual drying of single droplets on the substrate and prevents the formation of a closed liquid film. Hence, the surface roughness of spray coated films is typically in the tens of nanometers compared to significantly smoother spin coated or doctor bladed films, which are typically in the range of few nanometers.

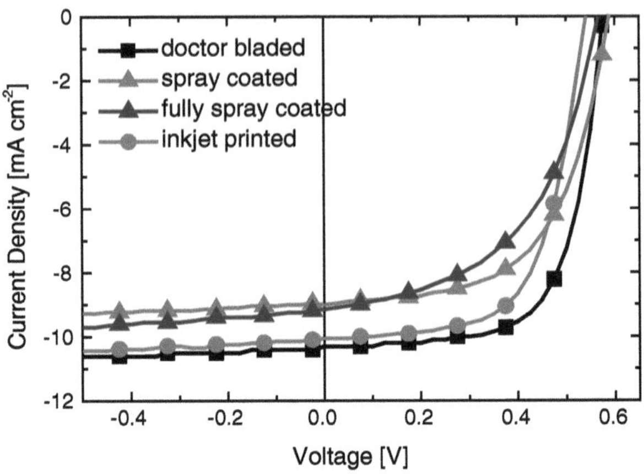

Fig. 2.6 Current–voltage behaviour of representative organic solar cells manufactured by doctor blading (*squares*) [64], spray coating (*triangles*) [45] and inkjet printing (*circles*) [63]. A fully spray coated solar cell device with spray coated PEDOT:PSS and spray deposited P3HT:PCBM is shown by the *dark triangles* [45]. Current density–voltage behaviour and device performance parameters of solar cell devices are studied with AM1.5G illumination with 100 mW cm^{-2}

To achieve rapid drying, the semiconductor ink must fulfil specific requirements, which are primarily defined by the solvent properties, such as boiling point, vapour pressure, viscosity and surface tension. Beyond this, the ink has to be compatible with the nozzle of the airbrush system: coating fluids for spray application need to be weakly viscous at ambient conditions since the inks are not heated during the spray deposition [45]. Highly viscous inks would require higher application temperatures to decrease the viscosity.

At constant nozzle-to-substrate distances, the liquid droplets of low drying-rate inks are immediately pushed sideward by the pressure gas of the airbrush. This leads to non-uniform wetting. On the other hand, spray coated droplets based on very quickly evaporating organic solvents may dry before reaching the substrate surface, preventing the formation of homogeneous films.

"Low wets high" is a guideline for printing. In fact low surface tension for the blended semiconductor solutions guarantees reasonable wetting on many substrates. For instance, PEDOT:PSS surfaces show high surface energies due to the hygroscopic behaviour of PEDOT:PSS and thus the P3HT:PCBM ink should have surface tensions in the 30 mN m^{-1} regime for adequate wetting [45]. Recently, two ink formulations for spray coating based on either pristine organic solvent chlorobenzene or on a two component solvent mixture comprising 68 % oDCB and 32 % mesitylene were studied. These inks have a different drying behaviour due to their vapour pressures [45]. More details on the development of the ink formulation can be found in [44]. Here an organic photovoltaic device based on the improved ink formulation using oDCB/mesitylene is discussed. Despite the high surface roughness, the spray deposited devices on glass based on oDCB/mesitylene show high device performance (Fig. 2.6—"spray coated" line), a V_{oc} of

588 mV, J_{sc} of 9.0 mA cm^{-2}, and a high *FF* of 0.59 corresponding to a PCE of 3.1 %.

This good photovoltaic performance proves the potential of spray coating as a deposition method for P3HT:PCBM blends. The *J–V* characteristics under illumination of spray coated devices compared to solar cells prepared by doctor blading or inkjet printing (which will be discussed in the next section) are depicted in Fig. 2.6. The spray coated solar cell shows reduced J_{SC} and *FF* compared to doctor bladed ones. The lower *FF* is attributed to increased series resistance R_S, whereas the reduced J_{SC} is related to different semiconductor film morphologies.

To further verify the potential of spray coating, the authors in [45] demonstrated a fully spray-deposited solar cell on glass comprising a spray coated PEDOT:PSS film together with a spray coated photoactive P3HT:PCBM layer, still resulting in a good device performance. The current–voltage behaviour under illumination is also illustrated in Fig. 2.6. The fully spray deposited solar cell has $V_{oc} = 560$ mV, $J_{sc} = 9.1$ mA cm^{-2} and $FF = 0.52$. This corresponds to a power conversion efficiency of 2.7 % [45]. The lower device performance of the fully sprayed solar cells compared to devices fabricated with doctor bladed PEDOT:PSS film and spray coated P3HT:PCBM dissolved in oDCB/mesitylene is attributed to the fact that the sprayed PEDOT:PSS film decreases the V_{oc} (5 % lower), and strongly worsens the *FF* (which is 13 % lower for the fully sprayed devices). The J_{SC} is comparable in the different devices. The decreased *FF* is related to increased surface roughness of the spray deposited PEDOT:PSS film (rms = 20.7 nm) and hence, to a poor interface [45]. Despite the higher surface roughness of spray coated films, this method might be a feasible technology to deposit multiple functional layers resulting in well performing photovoltaic devices. In addition, Hau et al. and Girotto et al. also reported on the use of spray coating to manufacture the electrode of OPVs using silver (Ag) nanoparticles dispersed in solution [46, 47]. These promising results on spraying all functional layers of organic solar cells prove the potential of spray coating as a convenient manufacturing method for large area and high throughput roll-to-roll production of solar cells.

2.2.5 Inkjet Printing

In the past few years, inkjet printing has evolved as an important patterning method, not only in the paper industry. Nowadays inkjet printing is probably the most widely used method for organic electronics. Inkjet printers can be divided into three categories: continuous inkjet, drop on demand (DOD) inkjet and bubble jet, all differing in the formation of droplets [48]. The DOD approach is commonly used for accurate deposition of inks on defined spots. This technology eliminates the need for post-patterning the coated films and allows a precise printing with micrometric resolution due to the formation of droplets in the picoliter regime. The use of inkjet printing has been extensively implemented in the field of organic field effect transistors [49–56] and organic light emitting diodes [49, 57–59], but only little work is found on inkjet printed organic photovoltaics [44, 60–64]. The application of the inkjet printing technology as a fabrication tool for organic devices is very promising due to the compatibility with various substrates. As shown in the schematic representation in Fig. 2.5b [44], the ink is transferred from a writing head to the substrate without direct contact with the surface [65]. A droplet is ejected at the nozzle orifice. As it hits the surface of the substrate, the droplet spreads and wets the substrate. Spreading and wetting depends on the surface tension of both the fluid and the substrate. Film formation occurs by the coalescence of adjacent droplets. A liquid bulk is built that shows particular drying kinetics controlled by the solvent properties. The viscosity needs to be in a range of 1–10 mPas. Higher viscosities up to 40 mPas can be printed with high frequencies by heatable ink reservoirs or temperature controlled print heads to decrease the viscosity for the droplet formation at the nozzle. The droplet formation at the jet and the jetting behaviour may be critical for higher viscosities. High concentrated solutions or particle dispersions may cause jetting problems which lead to nozzle clogging. This can be overcome by using appropriate solvent formulations which extend the ink shelf life.

The homogeneity of the inkjet printed films is, among other factors, determined by the substrate temperature. Most of the vendors offer tools equipped with a heated substrate holder to increase the drying rate of the inkjet printed film.

The vapour pressure of organic solvents plays an important role in the design of inkjet fluids. Since the nozzle orifice is open to the atmosphere when not printing, the solvent will evaporate. This causes an increase of the viscosity at the nozzle orifice. Consequently, the droplet characteristics in terms of velocity, volume and angularity (angle deviation) are affected. This phenomenon is more distinct for high vapour pressure solvents.

During the vaporization of low boiling solvents at the nozzle orifice, the solid blend materials may precipitate in the jet resulting in clogged nozzles [44].

The surface tension of the ink plays a major role in controlling the spreading and wetting on the substrate as well as the interaction with the printhead. A 10 picoliter droplet has a calculated diameter of 27 microns.

The ability of the ink formulation to give excellent droplet formation for photovoltaic applications is strongly dependent on the rheology of the polymer-fullerene solution and on the driving conditions of the print head. Inkjet printing requires significant efforts to properly design an ink formulation for reliable jetting and optimized film quality. In parallel, an optimized drive waveform is responsible for reliable jetting behaviour. A droplet ejected by the piezo-driven print head has a velocity of a few meters per second. Prior to the impact on the substrate, the droplet should afford a round shape avoiding tails and satellites. Such tails and satellites lead to serious problems in printing small structures, e.g. printing of materials emitting different colours in small pockets on tiles for OLED display applications or printing metal electrode grids on OPVs, where high resolution is a prerequisite for reasonable performance.

A variation in the droplet volume results in different film thickness. A method to adapt the droplet volume is to modulate the driving voltage. Higher voltages

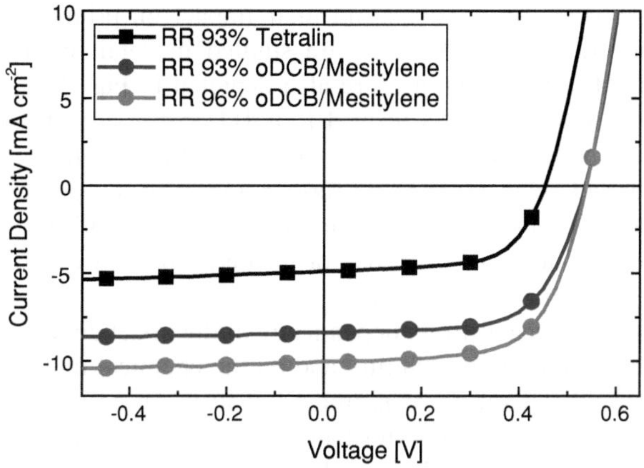

Fig. 2.7 Current–voltage behaviour of inkjet printed organic solar cells. The solvent formulation and the polymer material properties strongly impact the solar cell device performance [44, 63, 64]. Reproduced by permission of the Royal Society of Chemistry. Current density–voltage behaviour and device performance parameters of solar cell devices are studied with AM1.5G illumination with 100 mW cm^{-2}

result in larger droplets and thus, with equal droplet spacing, the printed film thickness can be increased. On top of this, film thickness can be controlled by modifying the droplet spacing or printing multiple drops per spot.

The investigations in [44] report on two ink formulations for OPVs differing in the drying behaviour due to their vapour pressures. These formulations are based on either pristine organic solvent tetrahydronaphthalene (tetralin) or on a two component solvent mixture comprising 68 % ortho-dichlorobenzene (oDCB) and 32 % 1,3,5-trimethylbenzene (mesitylene) [44]. Tetralin vaporizes and dries slower due to the lower vapour pressure (0.18 mmHg) and higher boiling point (207 °C). In contrast the mixture of oDCB and mesitylene results in homogeneous drying kinetics. Mesitylene features not only a higher vapour pressure (1.86 mmHg) and lower boiling point (165 °C) for rapid drying, but also a lower surface tension (28.8 dynes cm^{-1}). This allows better spreading and wetting behaviour with positive effects on the film formation and film quality. In [44] the authors investigate the morphological properties of the photoactive layer made with an inkjet printed polymer-fullerene blend and highlight the better device performance achieved using an optimized oDCB/mesitylene inkjet formulation instead of the pristine tetralin solvent. In this reference it is shown that solvent mixtures based on oDCB/mesitylene enhance the device performance of inkjet printed solar cells due to improved morphology, decreased first order recombination and reduced surface roughness. Accordingly, a calibrated power conversion efficiency of about 3 % was obtained for the new solvent choice (see Fig. 2.7—93 % oDCB/mesitylene trace) [44]. The main electrical performances of this device are: $V_{oc} = 540$ mV, $J_{sc} = 8.4$ mA cm^{-2} and $FF = 0.64$. The chemical material properties of the conjugated polymer such as regioregularity (RR) together with solvent formulations and their impact on printing reliability, film morphology and device performance are studied in detail in [63, 64]. The *J*–*V* curves obtained in that reference are also depicted in Fig. 2.7.

By using a combination of a 96 %-RR P3HT and a suitable solvent mixture oDCB/mesitylene to control the drying and film formation, the authors were able to achieve extended gelation time and further improved morphological properties of the active layer, resulting in a very high PCE of 3.5 % with V_{oc} of 537 mV, J_{sc} of 10.05 mA cm^{-2} and a *FF* of 0.64 for an inkjet printed conjugated polymer-fullerene bulk heterojunction solar cell.

These results evidence a strong dependency of the inkjet device performance on the solvent formulation, on its stability and on the chemical material properties of the donor materials. Therefore, it is important to make an effort on formulation work for high ink shelf life and performance, and adapt the ink to the printing tool. The development of suitable ink formulations for functional materials to control the morphology and printing reliability and achieve efficient devices is an essential step towards the commercial exploitation of organic photovoltaics.

2.2.6 Gravure Printing and Flexographic Printing

Gravure printing is one of the simplest printing technologies. The printing setup (Fig. 2.8a [66]) is based on an engraved printing roll, an ink bath, a coating knife and a counter pressure roll. The printing pattern is engraved into the printing cylinder, meaning that new patterns require new engraved rolls, leading to higher costs. The engraved cells are filled with ink by rotating the roll in an ink bath. The excessive ink is peeled off by a flexible coating knife (doctor blade) prior to contacting the substrate. Ink remains in the engraved pattern and is directly transferred onto the flexible substrate. During imprinting, the substrate takes up the ink from the engraved cells [66]. This method requires inks in the low viscosity regime. Gravure printing can result in high throughput and can be a cost-effective roll-to-roll printing technology. Gravure printing is capable of printing thin films in the range of 1–50 microns and patterns with line width smaller than 100 microns at high speeds up to 15 m/s [67]. These properties in combination with low temperature processing make this technology one of the most promising methods for the mass production of organic photovoltaics. The thickness of gravure printed films is proportional to the area and the depth of the engraved cells, and depends on the line speed, on the roll (cylinder) speed and on the nature of the ink in terms of concentration and rheological properties. Gravure printed organic solar cells were published in 2006 [66]. The authors reported excellent film qualities for 100 nm thick gravure printed P3HT:PCBM layers. A surface roughness of 1.58 nm rms was found for the gravure printed films, while spin coated films from the same solutions showed only slightly lower surface roughness values of 1.26 nm rms. Such a high quality print of thin films with low viscous inks is very encouraging. Beyond high film quality, the gravure printed photovoltaic device

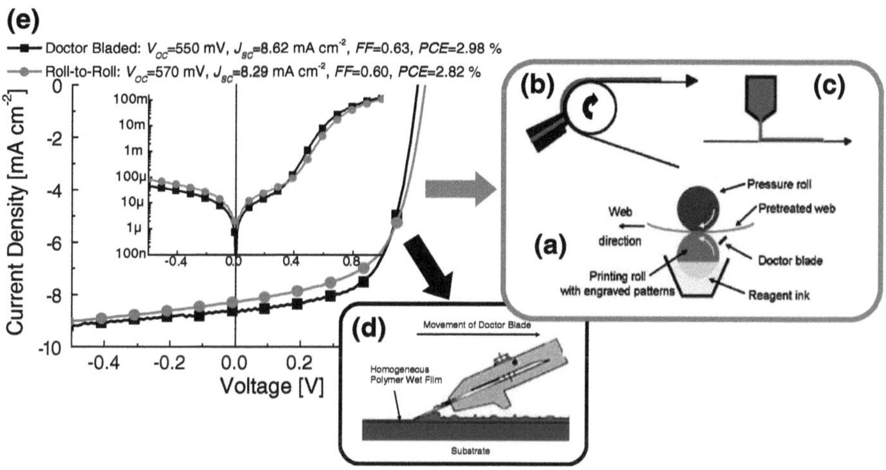

Fig. 2.8 Schematic representation of different fabrication technologies: **a** gravure printing [66], **b** slot die coating, **c** curtain coating [71] and **d** doctor blading. **e** Current–voltage behaviour of a lab-scale doctor bladed device (*black squares*) compared to a solar cell manufactured by a roll-to-roll process (*grey circles*). The *inset* shows the *J–V* characterization in the dark [Hoth et al. (2008) "unpublished"]. Current density–voltage behaviour and device performance parameters of solar cell devices are studied with AM1.5G illumination at 100 mW cm^{-2} intensity

showed good diode properties. The short circuit current density of these first gravure printed solar cell devices [66] was on the lower side for P3HT:PCBM composites, which might reflect the formation of an unfavourable morphology during the drying kinetics. These initial morphological problems seem to be overcome as Tuomikoski et al. recently demonstrated more than 2 % PCE for gravure printed solar cells under AM1.5G spectra. Similar encouraging results were obtained by Ding et al., who demonstrated printed flexible solar cell devices with a gravure printed hole-transporting layer as well as gravure printed photoactive layer (P3HT:PCBM). Power conversion efficiencies as high as 1.68 % under AM1.5G illumination have been reported for these fully printed solar cells [68]. These results demonstrate the feasibility of gravure printing for roll-to-roll manufacturing of organic solar cells. Gravure printing is able to meet all specifications with respect to layer homogeneity, surface roughness, resolution as well as production speed.

A printing method similar to gravure printing is flexographic printing, where the ink is transferred via flexible printing plates. A typical flexographic printing unit consists of 4 rolls: the fountain roll fills the Anilox roll (roll with engraved pattern). This roll feeds the printing roll with ink and the latter roll applies the ink to the substrate. The substrate is controlled by a back-up roll [69]. Flexographic printing plates lead to a broadening of the image resulting in reduced resolution and film quality. Flexographic printing allows the deposition of layers with comparatively high thicknesses up to several microns. Relatively low viscous inks can be handled at low transfer pressures.

The use of flexographic printing for organic electronic applications was discussed and demonstrated in great detail by Huebler et al. [70]. This group reported flexo-printed hole-transport layers as well as semiconductor layers with good homogeneity and sufficient resolution. However, so far, this printing method was used predominantly for organic electronic applications other than solar cells. Nevertheless, also flexographic printing has the potential to meet the production specifications for organic solar cells, similarly to gravure printing. Edge resolution may be lower than for gravure printing, but, organic photovoltaics won't need resolution better than a few hundreds of micrometers according to the first module designs.

2.2.7 Curtain Coating and Slot Die Coating

Curtain and slot die coating have evolved to promising printing methods for the manufacturing of organic solar cells using the roll-to-roll approach. In the curtain coating technique (Fig. 2.8c [71]), a gravity fed slot orifice is used to apply a coating by a falling ink curtain onto the substrate. The coating thickness depends on the die gap, the coating speed and certainly on the rheology of the coating fluid. Curtain coating can produce high quality films with a low number of physical defects at extremely high line speeds up to 1,000 m/min. Despite the importance of the volume flow density, which is of fundamental interest to achieve a stable curtain formation during coating, this method requires specifically designed ink formulations with high ink momentum to prevent defects. For the stability of the curtain, minimum line speeds are required.

In slot die coating inks are *squeezed* via a slot onto the substrate. In case of a 100 % solid coating, this process is called extrusion. An ink or coating is directly applied through a pump-fed slot orifice to a web supported by a backup roll, as shown in Fig. 2.8b. Both curtain and slot coating are rather comparable to doctor blading (Fig. 2.8d). Similar to doctor blading, the final film thickness in slot die coating depends not only on the nature of the ink, but on the coating parameters such as the die gap, the distance and angle between the die and the web, the pump speed, the web line speed and ultimately the temperature control of the flexible substrate. The application of slot die coating to either deposit the hole collecting PEDOT:PSS layer or the photoactive layer P3HT:PCBM were already discussed in [72, 73]. Recently, the first successful demonstration of fully slot-die-coated OPV cells was reported [74]. The functional layers were coated with a speed of 2 m/min and a coating width of 20–50 mm on ITO coated polyester film. An aluminium cathode was vacuum deposited onto small areas of the web coated film to complete the photovoltaic device resulting in power conversion efficiencies up to 1.74 % [74].

Slot die coating allows the usage of inks over a wide range of viscosities. However, the upper limit of the viscosities depends on the pump. The adjustment of the pump pressure is one critical parameter for slot die coating. At higher web speeds the material loss may increase, and the high costs of the die systems are another disadvantage. Nevertheless, the good homogeneity of coated films and the potential for high web speeds qualify slot die coating as a promising production technology for flexible organic photovoltaics [74]. Today's biggest drawback of slot die coating is its incompatibility with patterning. Production of OPV modules requires patterning the hole-transport as well as the semiconductor layer with sub-mm resolution and registration. Till today, it is not clear how to achieve this with classical coating methods. Post-patterning of thin organic layers, which was mentioned recently [75], might be a concept to allow exploring slot die coating for practical OPV module production.

2.2.8 Characterization of Proprietary Roll-to-Roll Printed Devices

To assess the quality of roll-to-roll processed films, our group investigated the impact of a proprietary roll-to-roll processing on the morphology and topography of roll-to-roll fabricated solar cells [Hoth et al. (2008) "unpublished"]. The device performance of a roll-to-roll deposited organic photovoltaic device compared to a sheet based doctor bladed solar cell fabricated at the laboratory scale is shown in Fig. 2.8e. Both types of solar cells had identical device architectures and were produced from identical materials onto an ITO coated polyester substrate. The *J–V* characteristics of the roll-to-roll processed solar cell have a V_{oc} of 570 mV, a J_{sc} of 8.29 mA cm^{-2}, a *FF* of 0.60 and a PCE of 2.82 %, all performances which are very similar to the doctor bladed device (V_{oc} of 550 mV, J_{sc} of 8.62 mA cm^{-2} and a *FF* of 0.63, corresponding to a PCE of 2.98 %). The inset in Fig. 2.8e represents the dark *J–V* curves. Both device types, despite being processed by different processing techniques, have similar values of the ideality factor *n* (1.6) and similar saturation currents J_0, indicating a very similar bulk morphology of the P3HT:PCBM composite. Accordingly, the domain size of the phase aggregation throughout the bulk must be of the same order. The lower V_{OC} of the doctor bladed solar cell is partly attributed to slightly higher injection currents over the roll-to-roll fabricated device. Furthermore, the decreased J_{SC} of the roll-to-roll device is explained by the slightly lower $\mu\tau$-product[4] compared to the doctor bladed reference solar cell. Nevertheless, the overall performance of the roll-to-roll processed organic solar cell is in good accordance with the lab-scale doctor bladed device and confirms the high quality of roll-to-roll processing as the method of choice for the production of OPV. Figure 2.9 [Hoth et al. (2008) "unpublished"] shows an overview of power conversion efficiencies achieved with different fabrication methods either on glass (black boxes) or flexible substrates (grey boxes).

In this section a selection of the most promising printing and coating methods for organic photovoltaic devices was investigated. Despite the differences of fluid

[4] The charge transport is field driven and the relevant transport parameters are the carrier mobility μ, carrier lifetime τ and internal electric field *E*.

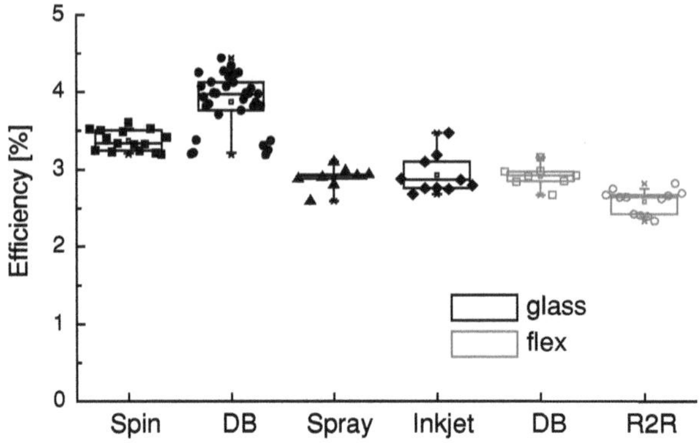

Fig. 2.9 Overview of power conversion efficiencies for different processing techniques for OPVs: spin coating (*Spin*) [26], doctor blading (*DB*) [26, 63, 64], spray coating (*Spray*) [45] and inkjet printing (*Inkjet*) [44, 63, 64], prepared on glass substrates (*black*). The *grey box plots* show doctor bladed (*DB*) and roll-to-roll processed (*R2R*) OPVs on flexible plastic carriers [Hoth et al. (2008) "unpublished"]. All solar cell devices were fabricated at the Konarka research and development facilities

handling and drying in the different methods and their different requirements with respect to ink formulation, we could show that all of the investigated technologies allow manufacturing cells and modules with nearly identical performance. This encourages to further develop new functional materials and ink formulations and to produce OPVs using large-scale printing or coating procedures.

2.3 Reliability and Yield of the Manufacturing Process

One fundamental requirement to a production technology which would be suitable for the market is a high degree of reliability and reproducibility of the results. To establish a baseline of efficiency and prove the reliability of standard manufacturing processes, the fabrication of OPVs by doctor blading was repeated multiple times. Figure 2.10 shows the efficiency evolution of a number of devices fabricated over several months under standard measurement conditions.

The baseline evidences the stable device performance of doctor bladed solar cells over several months. A standard sheet-based device process is thus extremely reliable and results in good-quality working devices with an observed yield reaching 90 %. The lower processing speed and the reduced throughput of the sheet based methods is not a drawback for laboratory operation, where a stable device performance baseline plays a major role for lifetime and degradation investigations and for any optimization approaches.

Similar yield levels close to 90 % are observed for roll-to-roll processing. The high yield of these processes is ascribed to the high degree of reliability of roll-to-roll technologies, and to the reproducibility and homogeneity of the functional films obtained [74].

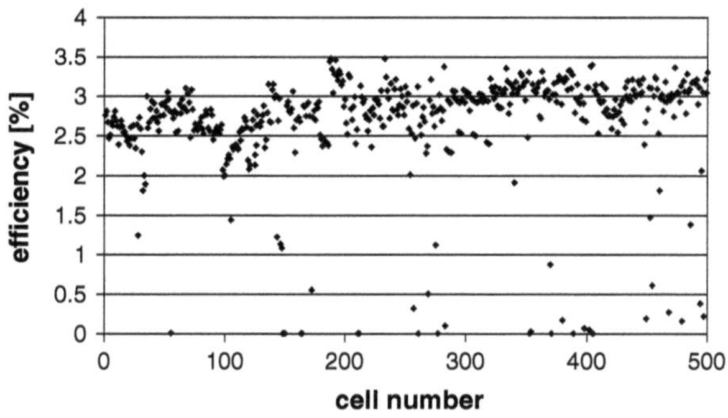

Fig. 2.10 Efficiency baseline shows the reproducibility of a sheet-fed coating process for flexible OPV cells. The efficiency is plotted vs. the number of manufactured cells

2.4 Accelerated Lifetime Test

The competitive position of OPVs in the energy market and their field of application is a function of efficiency, lifetime and in particular of cost per kWh. The potential of organic photovoltaics has to be assessed by these key parameters as well. This section will focus on lifetime and degradation of organic photovoltaics.

No matter the specific product application, lifetime characterization and understanding the degradation mechanisms is of great interest for a successful product. Degradation of solar cells is normally quantified by the time it takes for a given OPV to drop in power output to 80 % of the initial value [76, 77]. In the past, qualitative experiments to determine the lifetime showed that parameters such as temperature, humidity, electrical stress and irradiation are limiting the long-term stability and thus, accelerating the degradation of organic photovoltaic devices. To study degradation, obtain information about the parameters influencing lifetime, and to quickly characterize degradation mechanisms, accelerated lifetime testing is performed on organic solar cells which are exposed to stress conditions. Such stress conditions are realized in climate chambers with temperatures of 65 or 85 °C, in humid or dry atmosphere and exposing or not the device to 1 sun illumination (corresponding to a 100 mW/cm^2 intensity). Lifetime studies are also carried out with outdoor exposure. The determination of acceleration factors and, consequently, the use of accelerated lifetime models is an adequate instrument for predicting the lifetime of organic solar cells. The mechanism of the accelerated lifetime degradation at elevated temperatures and stress conditions can be described by an Arrhenius model [78]. Once the acceleration factor is determined at a certain stress condition, the lifetimes of the organic solar cell devices can be predicted.

Only limited data are available on the stability of organic solar cells [76, 77]. Hauch et al. reported on the stability of flexible organic solar cells, which were packaged with a low-cost and rather high water vapour transmission rate (WVTR) food-packaging barrier before exposing to 1 sun illumination at a temperature of 65 °C. In parallel, cells and modules with the same architecture and material combination were investigated at an outdoor rooftop setup in Lowell, MA (USA) to determine the lifetime in operative conditions [76]. The flexible polymer solar cells survived 1,000 h under accelerated light soaking conditions indicating a high light stability, with hero cells showing less than 20 % degradation after 6,000 h. The good environmental stability measured in accelerated experiments was confirmed by the outdoor data, where flexible modules located at the rooftop did not show performance losses after being operated for 1 year. Figure 2.11a shows the normalized power output (black diamonds) of a solar cell module exposed to outdoor conditions on the rooftop over a period of more than 1 year. The grey squares indicate the air temperature at the time of measurement. Immediately after positioning the module on the rooftop, the efficiency increased by 40 %. Afterwards, there is a strong correlation between efficiency and outdoor temperatures. In May 2007 a linear degradation trend sets in, resulting in a power output of 80 % of its initial value at the end of the experiment, after about 13 months. Re-measuring the *J*–*V* characteristics of the solar module after terminating the rooftop experiment indicates a slight increase in the power conversion efficiency due to a relative increase in *FF*, together with a slight decrease in V_{oc}, resulting in an overall slightly enhanced PCE. Due to the fact that the current density remains constant over the period of outdoor exposure, there is no hint for bleaching of the photoactive layer. The drop in the power output measured on the rooftop was thus only attributed to a shift in the maximum power point (MPP) of the module [76], which can be determined on *J*–*V* characteristics as indicated in Fig. 2.2b.

The impact of the WVTR of the packaging for flexible OPVs was also studied [77]. In this report, organic solar cells were encapsulated with a low-cost food package barrier film and exposed to three different climate conditions, 65 °C (high temperature storage), 65 °C illuminated with 1 sun (sun soaking) and 65 °C with 85 % relative humidity (damp heat). An abrupt 5 % drop in the V_{oc}, occurring within the first 100–200 h, is observed for the damp heat condition and may be attributed to a “soaking effect” of the device, i.e. the active layers are assumed to be saturated of humidity and in equilibrium with their surroundings. After a period where V_{OC} is stable in humid air until about 1,250 h, the cell rapidly fails [77]. The stability in V_{oc} hints to stable interfaces between the active layer and the electrode.

A strong degradation in J_{sc} is observed for the sun soaking conditions, while high temperature storage in dark and damp heat show better J_{sc} stability. The *FF* decreases constantly for all three conditions. The *FF* is more stable for the illuminated devices, while in dark *FF* degradation is more pronounced. The efficiency performance of the devices exposed to different climate conditions over time is plotted in Fig. 2.11b [77]. From these results it is clear that different kind of environmental stress lead to different types of degradation behaviour. Damp heat is the most damaging environmental condition when considering the time needed to drop to 80 % of the initial power output. Comparing the results of this study [77] with previously published data on accelerated lifetime measurements [76], the operational lifetime for flexible organic photovoltaic devices with an appropriate package under outdoor conditions can be predicted to be nowadays in the range of 3–5 years. Higher lifetimes are expected to come in quickly, either using intrinsically more stable materials or using packaging materials with better barrier properties.

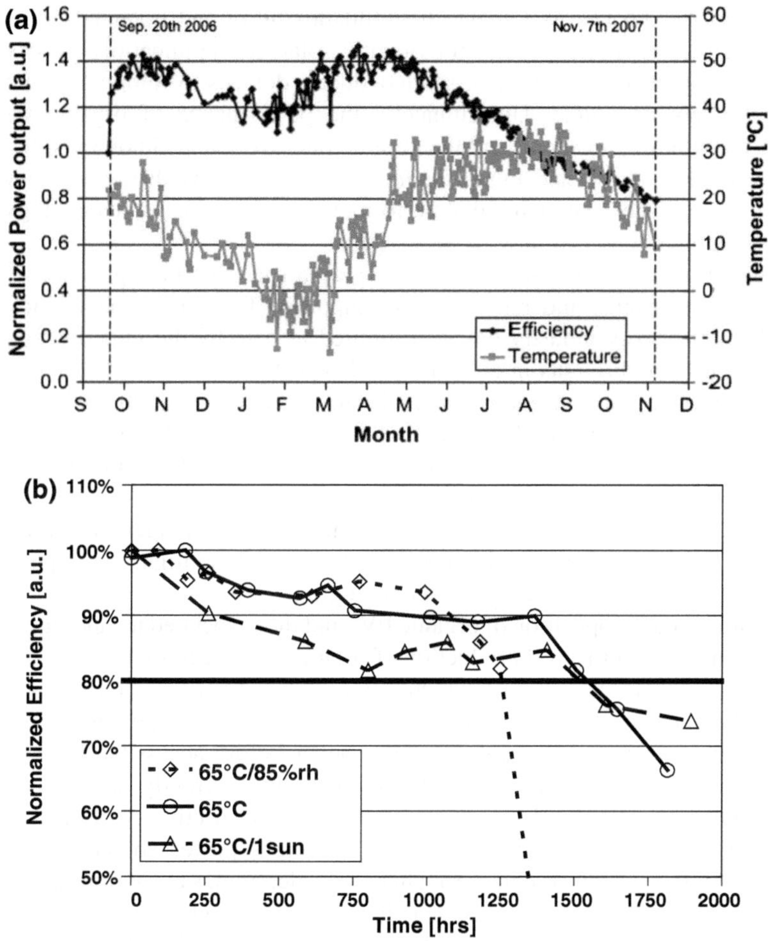

Fig. 2.11 Accelerated lifetime test. **a** Normalized power output (*black diamonds*) of a solar cell module on the rooftop over time in months. The *grey squares* indicate the air temperature at the time of measurement. Reproduced with permission by Elsevier from [76]. **b** Degradation in PCE (normalized) of flexible encapsulated organic solar cells exposed to three different climate conditions: 65 °C (*circles*), 65 °C illuminated with 1 sun (*triangles*) and 65 °C with 85 % relative humidity (*diamonds*). Reprinted with permission from [77]. Copyright 2008, American Institute of Physics

2.5 Application Aspects and Product Roadmap

In a market study [79], it has been predicted that organic semiconductors solar cells will have annual growth rates of 40 % or more after market introduction. This work also states that the "Development of high-efficiency organic solar cells would be the single most important factor in the emergence of ultra-portable applications like electronic textiles. Integrated solar cells would allow organic semiconductor devices to achieve their full flexibility and potential." Generally speaking, the short term vision for organic solar cell applications is to support market adoption of other organic semiconductor devices by supplying a portable, non-bulky power source. Unfortunately, the performance of organic semiconductor technology in general and the organic photovoltaic technology specifically, has been till recently inadequate for most applications. With the enormous progress made in recent years in OPV performance and production technology many of these applications can now be accessed.

Organic photovoltaics are expected to play a role also in applications beyond consumer electronics and remote power supplies. The scope and impact of the overall solar energy industry has been limited to date by the inability to achieve meaningful breakthroughs in efficiency and cost. In spite of this, the PV industry has been growing at a staggering annual rate of 44 % since 2002, with a total of just over 3 GW shipped in 2007. At an average price of $3/W, this represented in the same year a market of over $9 billion for modules and $18 billion for the total industry. After declining for 50 years, the cost of PV began to rise in 2003 due to a silicon feedstock shortage. These material issues present a longer term threat to the economic viability of conventional photovoltaics, and highlight the importance of developing non-silicon based solar technologies.

The first value proposition of organic PVs and the associated technology is large-scale continuous roll-to-roll production of solar modules achieving costs below $1/W peak, which is the price at which PV becomes truly competitive with grid-supplied electricity. OPVs' second value proposition is the flexible, conformable format, enabling integration into products never before accessible by PV, which would include, but not be limited, to anything that currently uses a battery for power. Further anticipated increase in efficiency and lifetime will enable manufacturing highly efficient flexible and light-weight power sources that can be integrated into products, ranging from handheld electronics to rooftops and facades.

Quality of life, environmental and health benefits related to a reduced consumption of fossil fuels are interrelated and will determine the market size of photovoltaics.

Due to its form factor, the colours and the rather high performance of semi-transparent cells, building integrated PVs (BIPV), and especially window integrated PVs, are one of the big opportunities for OPV. Assuming only 30 % of the US based existing housing stock is accessible for building and window integrated PV, and that 2 kW can be installed per structure, the longer-term building-integrated application would generate significant emissions reductions, with the potential for over 70 GW of power generation in the housing market alone. Assuming that 70 GW of PVs are installed by 2020 and produce power for 6 h (a rather optimistic guess) a day, 153 billion kWh per year would be generated. Using Annual Energy Outlook 2008 data, this represents 5 % of total projected US electricity consumption in 2020, and would eliminate 148 million tons of CO_2, 192,000 tons of SO_2, 108,000 tons of NO_x, and 990,000 tons of mercury emissions per year.

Sensors and indoor applications are a second key market for OPVs. Typical fire, smoke, and heat detectors used as safety devices in virtually all residences are an

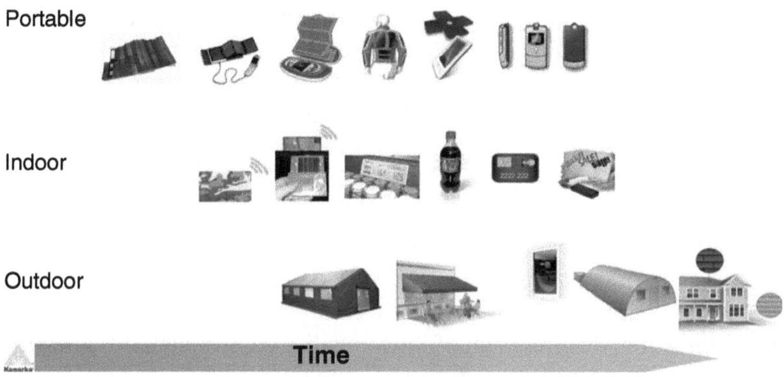

Fig. 2.12 Application aspects and product roadmap

example of such products. Despite the high level of market penetration, this is still a $550M annual market just for North America, driven by retrofits and new construction [80]. Today, these devices are either hardwired to the site electricity or battery powered. Many of these devices are in normally occupied areas of the residence, providing a relatively reliable source of ambient light. By implementing OPV panels (which were proven to function very well indoors as well as outdoors, unlike crystalline silicon) directly into the housing of the detector, the expense and inconvenience of hardwiring could be avoided, and the need for disposable batteries eliminated.

The rooftop market will be pursued last, as it is currently dominated by crystalline silicon technology with higher conversion efficiency, despite other limitations such as weight and fragility. Once the OPV technology achieves module efficiencies beyond 7 %, the technology is expected to be able to compete with crystalline technology in rooftop markets based on other consumer parameters, such as aesthetics, cost, ease of installation and weight [81].

Summarizing, the roadmap to market introduction that can be foreseen nowadays is shown in Fig. 2.12. The initial product focus is on the battery charging market, followed by consumer products (lighting, sensors, communications, computing), remote power applications (tents, awnings, shading elements) and finally building-integrated and grid connected PVs. This roadmap follows the increasing economy of scale from the pilot lines stage to the large-quantity production of OPV modules.

2.6 Conclusion

The encouraging results for printed organic solar cells strongly suggest that roll-to-roll fabrication is the most economical way to realize a mass production of photovoltaics. Organic photovoltaics, i.e. solar cells based on solution processable semiconductors, can be mass manufactured by sheet-fed and roll-to-roll printing/

coating technologies. In this chapter we have investigated in detail several printing and coating methods suitable for a P3HT: PCBM bulk heterojunction composite. Despite the differences of fluid handling and drying in the described methods and their different requirements with respect to ink formulation, we could show that all of the investigated technologies enable manufacturing cells and modules with nearly identical performance. An overview of these devices and their performance is shown in Fig. 2.9.

Significant technological developments have been reported on the printing and manufacturing of organic photovoltaics over the last years. Further progress is expected from continuously improved ink formulation, material development, process optimization, development of inline quality control methods and packaging methods. In parallel, with higher efficiencies and longer lifetimes OPV will become able to contribute to the world's energy supply, while first OPV products for consumer electronics, BIPV and remote power applications have entered the market starting in the last quarter of 2009 [82, 83].

References

1. Sirringhaus H, Brown PJ, Friend RH, Nielsen MM, Bechgaard K, Langeveld-Voss BMW, Spiering AJH, Janssen RAJ, Meijer EW, Herwig P, de Leeuw DM (1999) Two-dimensional charge transport in self-organized, high-mobility conjugated polymers. Nature 401:685
2. Brown AR, Pomp A, Hart CM, de Leeuw DM (1995) Logic Gates Made from Polymer Transistors and Their Use in Ring Oscillators. Science 270:972
3. Brabec CJ, Dyakonov V, Parisi J, Sariciftci NS (eds) (2003) Organic photovoltaics: concepts and realization. Springer series in materials science, vol 60. Springer, London
4. Gur I, Fromer NA, Geier ML, Alivisatos AP (2005) Air-stable all-inorganic nanocrystal solar cells processed from solution. Science 310(5474):462–465
5. Kapur V, Kemmerle R, Bansal A, Haber J, Schmitzberger J, Le P, Guevarra D, Kapur V, Stempien T (2008) Manufacturing of 'ink based' CIGS solar cells/modules. Conference record 33rd IEEE photovoltaic specialists conference, IEEE, Piscataway, NJ
6. Sang B, Adurodija F, Taylor M, Lim A, Taylor J, Chang Y, McWilliams S, Oswald R, Stanbery BJ, Van Hest M, Nekuda J, Miedaner A, Curtis C, Leisch J, Ginley D (2008) Low cost copper indium gallium selenide by the FASST® process. Conference record 33rd IEEE photovoltaic specialists conference, IEEE, Piscataway, NJ
7. Reyes-Reyes M, Kim K, Carroll DL (2005) High-efficiency photovoltaic devices based on annealed poly(3-hexylthiophene) and 1-(3-methoxycarbonyl)-propyl-1-phenyl-(6,6)C61 blends. Appl Phys Lett 87:083506
8. Ma W, Yang C, Gong X, Lee K, Heeger AJ (2005) Thermally Stable, Efficient Polymer Solar Cells with Nanoscale Control of the Interpenetrating Network Morphology. Adv Funct Mater 15:1617
9. Li G, Shrotriya V, Huang J, Yao Y, Moriarty T, Emery K, Yang Y (2005) High-efficiency solution processable polymer photovoltaic cells by self-organization of polymer blends. Nat Mater 4:864
10. Mühlbacher D, Scharber MC, Morana M, Zhu Z, Waller D, Gaudiana R, Brabec CJ (2006) High Photovoltaic Performance of a Low-Bandgap Polymer. Adv Mater 18:2884
11. Zhu Z, Waller D, Gaudiana R, Morana M, Mühlbacher D, Scharber MC, Brabec CJ (2007) Panchromatic Conjugated Polymers Containing Alternating Donor/Acceptor Units for Photovoltaic Applications. Macromolecules 40:1981

12. Peet J, Kim JY, Coates NE, Ma WL, Moses D, Heeger AJ, Bazan GC (2007) Efficiency enhancement in low-bandgap polymer solar cells by processing with alkanedithiols. Nat Mater 6:497–500
13. Soci C, Hwang IW, Moses D, Zhu Z, Waller D, Gaudiana R, Brabec CJ, Heeger AJ (2007) Photoconductivity of a Low-Bandgap Conjugated Polymer. Adv Funct Mater 17:632
14. Morana M, Wegscheider M, Bonanni A, Kopidakis N, Shaheen S, Scharber MC, Zhu Z, Waller D, Gaudiana R, Brabec CJ (2008) Bipolar charge transport in PCPDTBT-PCBM bulk-heterojunctions for photovoltaic applications. Adv Funct Mater 18:1757–1766
15. NREL certificate Konarka, 8.29 % PCE (thickness > 200 nm, device area 1.031 cm^2) under ASTM G173 global spectrum, 17 November 2010
16. Blayo A, Pineaux B (2005) Printing Processes and their Potential for RFID Printing. Joint sOc-EUSAI conference, Grenoble
17. Konarka technologies, hompage www.konarka.com
18. Yu G, Heeger AJ (1995) Charge separation and photovoltaic conversion in polymer composites with internaldonor/acceptor heterojunctions. J Appl Phys 78:4510
19. Yu G, Gao J, Hummelen JC, Wudl F, Heeger AJ (1995) Polymer Photovoltaic Cells: Enhanced Efficiencies via a Network of Internal Donor-Acceptor Heterojunctions. Science 270:1789
20. Shaheen SE, Brabec CJ, Sariciftci NS, Padinger F, Fromherz T, Hummelen JC (2001) 2.5% efficient organic plastic solar cells. Appl Phys Lett 78:841
21. Padinger F, Rittberger RS, Sariciftci NS (2003) Effects of postproduction treatment on plastic solar cells. Adv Funct Mater 13:85
22. Ma W, Yang C, Gong X, Lee K, Heeger AJ (2005) Thermally Stable, Efficient Polymer Solar Cells with Nanoscale Control of theInterpenetrating Network Morphology. Adv Funct Mater 15:1617
23. Konarka homepage (www.konarka.com), press release 19 May 2009, NREL certificate 6.4 % PCE
24. Brabec CJ, Padinger F, Hummelen JC, Janssen RA, Sariciftci NS (1999) Realization of Large Area Flexible Plastic Solar Cells Based on Conjugated Polymers and Fullerenes. Synth Metals 102:861
25. Padinger F, Brabec CJ, Fromherz T, Hummelen JC, Sariciftci NS (2000) Fabrication of Large Area Photovoltaic Devices Containing various blends of Polymer & Fullerene Derviatives by Using the Doctor Blade Technique. Optoelectron Rev 8(4):280
26. Schilinsky P, Waldauf C, Brabec CJ (2006) Performance analysis of printed organic solar cells. Adv Funct Mater 16:1669
27. Chang Y-H, Tseng S-R, Chen C-Y, Meng H-F, Chen E-C, Horng S-F, Hsu C-S (2009) Polymer solar cell by blade coating. Org Electron 10(5):741–746
28. Waldauf C, Schilinsky P, Hauch JA, Brabec CJ (2004) Material and device concepts for organic photovoltaics: towards competitive efficiencies. Thin Solid Films 451–452:503–507
29. Schilinsky P, Waldauf C, Hauch JA, Brabec CJ (2004) Simulation of light intensity dependent current characteristics of polymer solar cells. J Appl Phys 95:5
30. Waldauf C, Scharber MC, Schilinsky P, Hauch JA, Brabec CJ (2006) Physics of organic bulk heterojunction devices for photovoltaic applications. J Appl Phys 99:104503
31. Shaheen SE, Radspinner R, Peyghambarian N, Jabbour GE (2001) Fabrication of bulk heterojunction plastic solar cells by screen printing. Appl Phys Lett 79:2996
32. Krebs FC, Alstrup J, Spanggaard H, Larsen K, Kold E (2004) Production of large-area polymer solar cells by industrial silk screen printing, lifetime considerations and lamination with polyethyleneterephthalate. Sol Energy Mater Sol Cells 83:293
33. Krebs FC, Spanggard H, Kjaer T, Biancardo M, Alstrup J (2007) Large area plastic solar cell modules. Mater Sci Eng B 138:106
34. Aernouts T, Vanlaeke P, Poortmans J, Heremans P (2005) Polymer solar cells: screen-printing as a novel deposition technique. Mater Res Soc Symp Proc 836, art. no. L3.9, 81

35. Aernouts T, Vanlaeke P, Geens W, Poortmans J, Heremans P, Borghs S, Mertens R, Andriessen R, Leenders L (2004) Printable anodes for flexible organic solar cell modules. Thin Solid Films 451–452:22
36. Krebs FC (2008) Air stable polymer photovoltaics based on a process free from vacuum steps and fullerenes. Sol Energy Mater Sol Cells 92:715–726
37. Ishikawa T, Nakamura M, Fujita K, Tsutsui T (2004) Preparation of organic bulk heterojunction photovoltaic cells by evaporative spray deposition from ultradilute solution. Appl Phys Lett 84:2424
38. Mo XL, Mizokuro T, Mochizuki H, Tanigaki N, Hiraga T (2005) Polymer Solar Cell Prepared by a Novel Vacuum Spray Method. Jpn J Appl Phys, Part 1, 44: 656
39. Vak D, Kim S, Jo J, Oh S, Na S, Kim J, Kim D (2007) Fabrication of organic bulk heterojunction solar cells by a spray deposition method for low-cost power generation. Appl Phys Lett 91:081102
40. Green R, Morfa A, Ferguson AJ, Kopidakis N, Rumbles G, Shaheen SE (2008) Performance of bulk heterojunction photovoltaic devices prepared by airbrush spray deposition. Appl Phys Lett 92:033301
41. Steirer KX, Reese MO, Rupert BL, Kopidakis N, Olson DC, Collins RT, Ginley DS (2009) Ultrasonic Spray Deposition for Production of Organic Solar Cells. Sol Energy Mater Sol Cells 93:447–453
42. Girotto C, Rand BP, Genoe J, Heremans P (2009) Exploring spray coating as a deposition technique for the fabrication of solution-processed solar cells. Sol Energy Mater Sol Cells 93:454–458
43. Hoth C (2009) Ink Formulations for Organic Photovoltaics and their Processing with Printing and Coating Technologies. Carl von Ossietzky Universität Oldenburg, Fakultät V, EHF, PhD thesis
44. Hoth CN, Choulis SA, Schilinsky P, Brabec CJ (2007) High Photovoltaic Performance of Inkjet Printed Polymer:Fullerene Blends. Adv Mater 19:3973
45. Hoth CN, Steim R, Schilinsky P, Choulis SA, Tedde SF, Hayden O, Brabec CJ (2009) Topographical and morphological aspects of spray coated organic photovoltaics. Org Electron 10:587–593
46. Hau SK, Yip H-L, Leong K, Jen AK-Y (2009) Spray coating of silver nanoparticle electrodes for inverted polymersolar cells. Org Electron 10:719–723
47. Girotto C, Rand BP, Steudel S, Genoe J, Heremans P (2009) Nanoparticle-based, spray-coated silver top contacts for efficient polymer solar cells. Org Electron 10:735–740
48. Gamota DR, Brazis P, Kalyanasundaram K, Zhang J (2004) Printed Organic and Molecular Electronics. Kluwer Academic Publishers, T. Claypole, pp 320–322
49. Calvert P (2001) Inkjet Printing for Materials and Devices. Chem Mater 13:3299–3305
50. Liu Y, Varahramyan K, Cui T (2005) Low-Voltage All-Polymer Field-Effect Transistor Fabricated Using an Inkjet PrintingTechnique. Macromol Rapid Commun 26:1955–1959
51. Kim D, Jeong S, Lee S, Park BK, Moon J (2007) Organic thin film transistor using silver electrodes by the ink-jet printing technology. Thin Solid Films 515:7692–7696
52. Song DH, Choi MH, Kim JY, Jang J (2007) Process optimization of organic thin-film transistor by ink-jet printing of DH4T on plastic. Appl Phys Lett 90:053504
53. Mannerbro R, Ranlöf M, Robinson N, Forchheimer R (2008) Inkjet printed electrochemical organic electronics. Synth Metals 158:556–560
54. Sirringhaus H, Kawase T, Friend RH, Shimoda T, Inbasekaran M, Wu W, Woo EP (2000) High-Resolution Inkjet Printing of All-Polymer Transistor Circuits. Science 290:2123
55. Kawase T, Sirringhaus H, Friend RH, Shimoda T (2001) Inkjet Printed Via-Hole Interconnections and Resistors for All-Polymer Transistor Circuits. Adv Mater 13:1601
56. Kawase T, Shimoda T, Newsome C, Sirringhaus H, Friend RH (2003) Inkjet printing of polymer thin film transistors. Thin Solid Films 438:279–287
57. Xia Y, Friend RH (2006) Polymer bilayer structure via inkjet printing. Appl Phys Lett 88:163508

58. Xia Y, Friend RH (2005) Controlled phase separation of polyfluorene blends via inkjet printing. Macromolecules 38:6466–6471
59. Shimoda T, Morii K, Seki S, Kiguchi H (2003) Inkjet Printing of Light-Emitting Polymer Displays. MRS Bull pp 821–827
60. Shah VG, Wallace DB (2004) Low-cost Solar Cell Fabrication by Drop-on-Demand Ink-jet Printing. Proceedings of IMAPS 37th annual international symposium on microelectronics, Long Beach, CA, 14–18 Nov 2004, pp 1
61. Marin V, Holder E, Wienk MM, Tekin E, Kozodaev D, Schubert US (2005) Ink-Jet Printing of Electron Donor/Acceptor Blends: Towards Bulk Heterojunction SolarCells. Macromol Rapid Commun 26:319–324
62. Aernouts T, Aleksandrov T, Girotto C, Genoe J, Poortmans J (2008) Polymer based organic solar cells using ink-jet printed active layers. Appl Phys Lett 92:033306
63. Hoth CN, Schilinsky P, Choulis SA, Brabec CJ (2008) Printing highly efficient organic solar cells. Nano Lett 8:2806–2813
64. Hoth CN, Choulis SA, Schilinsky P, Brabec CJ (2009) On the effect of poly(3-hexylthiophene) regioregularity on inkjet printed organic solar cells. J Mater Chem 19(30):5398–5405
65. Nie Z, Kumacheva E (2008) Patterning surfaces with functional polymers. Nature Mater p 7
66. Tuomikoski M, Suhonen R (2006) Technologies for Polymer Electronics. Proceedings of TPE06, 2nd international symposium technologies for polymer electronics, Rudolstadt, vol 83
67. Yin X, Kumar S (2006) Flow visualization of the liquid-emptying process in scaled-up gravure grooves and cells. Chem Eng Sci 61:1142–1152
68. Ding JM, De la Fuente Vornbrock A, Ting C, Subramanian V (2009) Patternable polymer bulk heterojunction photovoltaic cells on plastic by rotogravure printing. Sol Energy Mater Sol Cells 93:459–464
69. Krebs FC (2009) Fabrication and processing of polymer solar cells: A review of printing and coating techniques. Sol Energy Mater Sol Cells 93:394–412
70. Hübler AC, Kempa H (2008) Flexo printing in organic electronics. In: Brabec CJ, Dyakonov V, Scherf U (eds) Organic photovoltaics. Wiley VCH, New York
71. Santurri P, Chemsultants, Inc., (2007) Coating methods for producing polymer films & membranes 3rd MEA Manufacturing symposium, Dayton, Ohio
72. Schultheis K, Blankenburg L, Sensfuss S, Schrödner M (2007) Polymer photo-voltaics: first steps to large scale R2R-production using wet coating techniques. Proceedings of the Cintelliq conference, Frankfurt
73. Schrödner M, Schultheis K, Blankenburg L, Schache H, Sensfuss S (2008) Reel-to-reel film coating technique for production of functional layers for polymer photovoltaics and electronics. International symposium TPE08, Rudolstadt
74. Blankenburg L, Schultheis K, Schache H, Sensfuss S, Schrödner M (2009) Reel-to-reel wet coating as an efficient up-scaling technique for the production of bulkheterojunctionpolymer solar cells. Sol Energy Mater Sol Cells 93:476–483
75. Tipnis R, Bernkopf J, Jia S, Krieg J, Li S, Storch M, Laird D (2009) Large-area organic photovoltaic module - fabrication and performance. Sol Energy Mater Sol Cells 93:442–446
76. Hauch JA, Schilinsky P, Choulis SA, Childers R, Biele M, Brabec CJ (2008) Flexible organic P3HT:PCBM bulk-heterojunction modules with more than 1 yearoutdoor lifetime. Sol Energy Mater Sol Cells 92:727–731
77. Hauch JA, Schilinsky P, Choulis SA, Rajoelson S, Brabec CJ (2008) The impact of water vapor transmission rate on the lifetime of flexible polymer solar cells. Appl Phys Lett 93:103306
78. Schuller S, Schilinsky P, Hauch JA, Brabec CJ (2004) Determination of the degradation constant of bulk heterojunction solar cells by accelerated lifetime measurements. Appl Phys A 79:37
79. Derbyshire K (2003) The future of organic semiconductors. Pira International, Surrey

80. Frost & Sullivan (2004) Security technology—North American trends and developments in video surveillance D288
81. Dennler G, Brabec CJ (2008) In: Brabec CJ, Dyakonov V, Scherf U (eds) Organic photovoltaics: Materials, device physics & manufacturing technologies. Organic photovoltaics. Wiley VCH
82. Press release, Konarka Homepage (www.konarka.com), Eröffnung der New Bedford Produktionsanlage, 2008
83. Konarka homepage (www.konarka.com), power plastics, products and applications

Chapter 3
High-Performance Organic Light-Emitting Diode Displays

Jang Hyuk Kwon, Ramchandra Pode, Hye Dong Kim and Ho Kyoon Chung

Abstract The development of new display devices for the interactive communication between computers and people has accelerated over the past decade and considerable progress has recently been made in the area of organic displays. Organic light-emitting devices (OLEDs) are the most suitable candidate to satisfy the demands of next generation displays among the technological options available so far, owing to simple device configuration, high power efficiency and efficient driving schemes, together with solid state encapsulation and excellent user experience. Efficient OLED structures, processes for OLED fabrication, various driving schemes for OLED displays, the current status of fluorescent and phosphorescent OLEDs, top emitting active matrix OLED (AMOLED), passive matrix driving schemes, white OLEDs for high resolution display applications, and thin film transistor (TFT) backplane technology for active matrix OLEDs are discussed here. Finally, the future scope and directions of the high-performance OLED display in mobile display technology and large area TVs are presented.

Keywords Display · OLED · Active matrix OLED · Thin film transistor · Pixel patterning · Phosphorescence

---

J. H. Kwon (✉)
Department of Information Display, Kyung Hee University, Dongdaemoon-ku, Seoul 130-701, Korea
e-mail: jhkwon@khu.ac.kr

R. Pode
Department of Physics, Kyung Hee University, Dongdaemoon-ku, Seoul 130-701, Korea

H. D. Kim · H. K. Chung
Samsung Mobile Display Co. Ltd, San 24, Nongseo-dong, Giheung-gu, Yongin City, Gyenggi-do, Korea

E. Cantatore (ed.), *Applications of Organic and Printed Electronics*, Integrated Circuits and Systems, DOI: 10.1007/978-1-4614-3160-2_3,

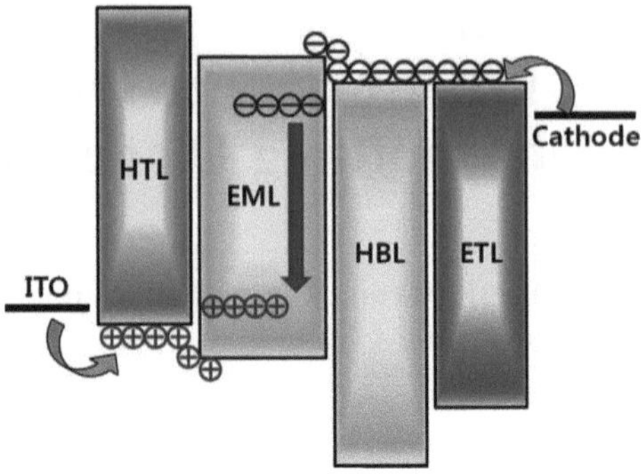

Fig. 3.1 Block diagram of OLED device where *ITO* (Indiun tin oxide) is the anode, *HTL* is the hole transport layer, *EML* is the emitting layer, *HBL* is the hole blocking layer, *ETL* is the electron transport layer and the metal cathode is calcium, magnesium, or lithium fluoride-aluminum

3.1 Introduction

Communication of information today is highly visual and electronic displays are the primary means of interactive communication between computers and people. High-performance information display devices are thus increasingly important due to remarkable advancements in communication technology [1, 2].

The development of new display devices has accelerated over the past decade and considerable progress has recently been made in the area of organic displays which meet the key requirements for wide-spread consumer adoption [3, 4]. Essential requirements for next generation displays are high-quality video image reproduction, high brightness and contrast, improved color variation and resolution, light weight, thin flat panels, and low power consumption. To satisfy the demands of next generation displays, organic light-emitting devices (OLEDs) are the most suitable candidate among the technological options available so far owing to simple device configuration, high power efficiency and efficient driving schemes, together with solid state encapsulation.

OLEDs are heterostructure devices, generally consisting of an organic emitting layer sandwiched between a transparent indium tin oxide (ITO) anode and a low work function cathode fabricated using calcium, magnesium, or lithium fluoride-aluminum as displayed in Fig. 3.1. In this chapter, we discuss efficient OLED structures, processes for OLED fabrication, and various driving schemes for OLED displays. The importance of the topic is discussed in this section. The evolution of OLED devices, the current status of fluorescent OLED technology and of phosphorescent OLEDs required for high brightness applications are discussed in Sects. 3.2, 3.3, and 3.4, respectively. Top emitting active matrix OLED (AMOLED), which have advantages over the passive matrix driving schemes, bottom emitting OLEDs and white OLEDs for high resolution display applications are discussed in Sect. 3.5. Thin film transistor (TFT) backplane technology for active matrix OLEDs is overviewed in Sect. 3.6. Finally, the future scope and directions of high-performance OLED display technology are illustrated in the Sect. 3.7 of this chapter.

3.2 Organic Light Emitting Devices

After the first report of electroluminescence in anthracene organic materials in monolayer devices in 1965 by Helfrich and Schneider [5], this phenomenon remained of pure academic interest for two decades, owing to the difficulty of growing large-size single crystals and the requirement of a very high voltage ($\sim$1,000 V) to produce the luminance. The evolution of OLED devices is summarized in Fig. 3.2. The poor performance of these early devices could be attributed to three main factors, namely (1) disparity between the mobility of electrons and the holes, (2) poor film-forming properties of the materials, and (3) mismatch between the energy levels of the organic materials and the contact electrodes. Tang [6] and Tang and VanSlyke [7] demonstrated that the poor performance of the early monolayer devices was dramatically improved in a two-layer device simply by the addition of a hole transport layer (HTL) in the device structure (Fig. 3.2). Organic electroluminescent devices having improved power conversion efficiencies by doping the emitting layer were also realized around the same time by the Kodak group [8, 9]. Subsequently, heterostructure configurations to improve the device performance were implemented by inserting several layers in the device structure, like a buffer film between anode and hole transport layer (HTL) [10–12], an electron transport layer (ETL), a hole blocking layer (HBL) [13] or an interlayer between cathode and ETL [14, 15]. Such multilayer structures often resulted in lowering of the driving voltage needed for the OLEDs. Usually, the operating voltage for high brightness was much higher than the thermodynamic limit, for example 2.4 eV for a green device. Chemical doping with either electron donors (for electron transport materials) or electron acceptors (for hole transport materials) can significantly reduce the voltage drop across these films [16–19]. These devices with either HTL or ETL doped layer showed improved performance; but the operating voltages were still higher than the thermodynamic limit. Later, Leo and his group proposed the concept of p-type doped HTL and n-type doped ETL [20]. These PIN-structure devices show high luminance and efficiency at extremely low operating voltages [21, 22].

3.3 Current Status of OLED Technology

As shown in Fig. 3.3, basic OLEDs have multilayer structures consisting of a hole injection layer (HIL), a HTL, an emissive layer (EML), a HBL (omitted in the figure), and an ETL, sandwiched between low-work-function cathode and transparent anode [15, 20]. These multiple layers are used to improve the brightness, the energy barrier at the electrode–organic interfaces, lifetime, and efficiency of the OLED. When an electric field is applied between the anode and cathode, electrons and holes are injected from cathode and anode, respectively, into the organic layers. With a matched energy barrier between the electron and hole

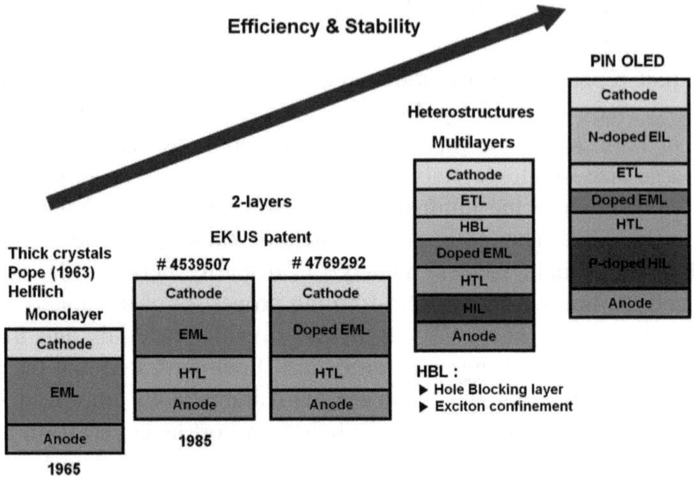

Fig. 3.2 The evolution of OLED devices. (*HIL* = hole injection layer, *HTL* = hole transport layer, *EML* = emissive layer, *HBL* = hole blocking layer, *ETL* = electron transport layer)

Fig. 3.3 Device structures of bottom and top emission *OLEDs*. (*HIL* = hole injection layer, *HTL* = hole transport layer, *EML* = emissive layer, *ETL* = electron transport layer)

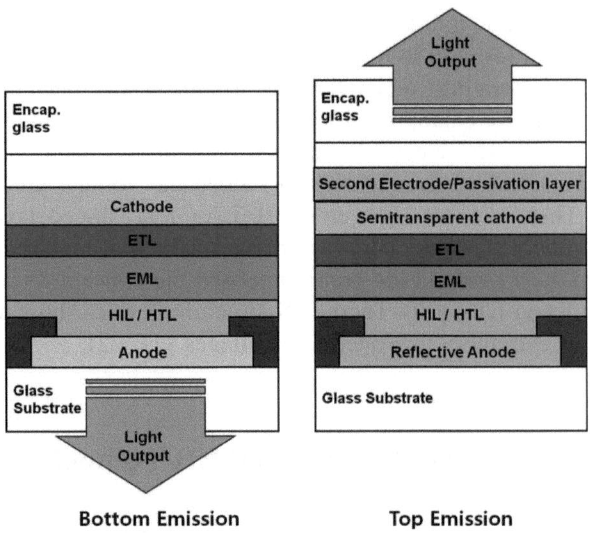

injection layers (EILs and HILs) and the cathode and anode, electrons and holes are efficiently injected into the ETL and HTL. Once the electrons and holes have been injected, they migrate into ETL and HTL, which are characterized by good charge mobility. The charges then meet in the organic EML and the device is optimized by fine-tuning so as to match the number of electrons and holes coming through the HTL or HBL (Fig. 3.1). Once the opposite charges recombine, an exciton is formed. Depending upon the nature of the spin state, singlet fluorescence[1] or triplet phosphorescence[2] light is emitted.

Recently, substantial progress has been made in the rapidly emerging field of OLED technology. The device efficiency and lifetime have been dramatically improved over the last 10 years. As a result of excellent progress in OLED materials and device technologies, operational lifetime is no longer an issue for the mobile applications. The lifetime of OLEDs is closely related to the material performances of their organic layers. Hence, the charge transport and light-emitting layers are required to have a high thermal and electrochemical stability as long term OLED operation leads to crystallization and electrochemical degradation.

Controlling the charge balance [13] in organic layers, i.e. controlling charge injection into the emitting layer from anode to cathode, is another important factor when trying to improve the device lifetime. Charge imbalance in OLEDs causes current leakage and charge build-up in the device, which leads to luminescence fading and a voltage rise during operation [23–25]. The electron and hole balance in OLEDs is essential not only to achieve a good lifetime, but also for improved efficiency and power consumption. Moreover, highly efficient OLEDs are expected to have better stability as non-radiative processes are less likely to occur.

On top of the issues discussed above, the interfacial contact between organic layers and electrodes, cathode structure, and interlayer diffusion influence the lifetime of OLEDs [26]. The current state of the art for lifetime of fluorescent OLEDs and phosphorescent OLEDs (PHOLEDs) is summarized in Table 3.1 [23–25]. PHOLEDs, which use both singlet and triplet states for light emission should have longer lifetimes than fluorescent OLEDs. However, PHOLEDs reported in the literature [23–25] have a relatively shorter lifetime than fluorescent OLEDs. The longer exciton lifetime observed in PHOLEDs may contribute to poor device stability. The red and green PHOLEDs show reasonable lifetime, while blue PHOLEDs are still under development for commercial applications. Although a significant improvement in OLED lifetime has been realized over the last 10 years, the operational lifetimes need to be further improved for large screen TV applications. The lifetime of RGB colors should be greater than 50,000 h while producing white light at 1,000 cd/m^2 in AMOLED displays. In general, to reach 1,000 cd/m^2 white brightness in an AMOLED display each red, green, and blue pixel should reach 3,500, 7,000, and 1,500 cd/m^2 brightness respectively, owing to factors like the low aperture ratio of the display panel and the transmittance of the contrast enhancement film.

---

[1] Fluorescence: Some materials absorb light at a particular wavelength and then emit light at a longer wavelength (lower frequency) than the incident light without changing the direction of the *e-* spin. In general this effect appears in a ten nanosecond timescale.

[2] Phosphorescence: is similar to fluorescence. Both involve absorption of energy. The light emission process must wait until the electron undergoes a spin-flip to revert back to its original state. Phosphorescence has much longer timescale (generally 10^{-3}–10^{2} s) and is controlled by "forbidden" quantum physics transitions.

Table 3.1 Performance overview of bottom emitting OLEDs

Color	CIE 1931 color coordinates	External quantum efficiency (%)	Luminous efficiency (cd/A)	Operational lifetime to 50 % (h)	Initial luminance (cd/m^2)	Voltage (V)
Fluorescent Red	(0.66, 0.34)	8.5	10	>500,000	1,000	2.5
Fluorescent Green	(0.35, 0.60)	8.5	32	>500,000	1,000	2.5
Fluorescent Blue	(0.14, 0.15)	5.0	6.1	30,000	1,000	3.5
Phosphorescent Deep Red	(0.67, 0.33)	21	21	80,000	1,000	4.3
Phosphorescent Red	(0.64, 0.36)	21	28	500,000	1,000	2.8
Phosphorescent Green	(0.36, 0.60)	16	58	100,000	1,000	2.8
	(0.38, 0.59)	19	67	250,000	1,000	4.6

Sample performance data in bottom emission structures (with no cavities)
The operational lifetime data is based on accelerated current drive conditions at room temperature
The lifetime of blue phosphorescence OLEDs in [25] is under 5,000 h and still not sufficient for commercial use
CIE = Commission Internationale de l'Eclairage

In order to implement large screen TV applications, the remaining key problem is the lifetime of blue-emitter devices. The longest reported lifetime of fluorescent blue devices is 30,000 h of continuous operation until the luminance of the OLED drops to half its initial brightness (1,000 cd/m^2) [23]. This value is consistent with the current AMOLED mobile applications which have lifetimes between 30,000 and 60,000 h at an approximate brightness of 200 $\sim$ 300 cd/m^2 [27], but the lifetime of blue emitters must still improve for use in large screen TVs. Nowadays, in full color commercial displays, blue and green fluorescent OLEDs and red phosphorescent OLEDs are employed together to achieve at the same time high efficiency and satisfactory product lifetime [23, 24].

3.4 Efficient Phosphorescent OLED Devices

In recent years, PHOLEDs are gaining dominant position in the field of OLED devices owing to their superior efficiency, which makes them suitable to high-performance high-brightness displays and to solid state lighting [28–30]. The upper limit in external quantum efficiency of about 5 % observed in fluorescent small molecule organic devices has been overcome by harvesting both the singlet and the triplet excitons to produce a large emission of photons in PHOLEDs [31, 32]. Iridium (III) and platinum (II) complexes are well-known phosphorescent emitters. The iridium (III) complexes have been shown to be the most efficient triplet dopants for high efficiency OLEDs [33, 34]. To produce high quantum efficiency in phosphorescent OLEDs, the excited energy of the phosphorescent emitter has to be confined within the emitter itself. This has been achieved using multilayer device architectures comprising electron/hole injection and transport layers with wide energy gap host and carrier-transporting materials [35, 36].

Due to the multilayer structure needed to assure good charge balance and to confine excitons within the emitting material (EML) [33, 35], the turn-on voltage of conventional PHOLEDs was 1–2 V higher compared to that of fluorescent OLEDs. This was an important limiting factor in the use of PHOLEDs for commercial display applications. Usually, wide energy gap materials (~3.0 eV) had been used as the host for red (~2.0 eV) or green (~2.3–2.4 eV) phosphorescent guests [37, 38]. Such wide energy gap hosts have high singlet–triplet exchange energies of >0.5 eV. There is a significant difference in the HOMO and/or LUMO levels between the guest and host materials for wide band gap host and narrow band gap guest system [38]. Thus, the guest molecules are thought to act as deep traps for electrons and holes in the emitting layer, causing an increase in the drive voltage of the PHOLED [39]. The dopant concentration in such a host-guest system is usually as high as about 6 ~ 10 % by weight (wt %) because the injected charges move through the dopant molecules in the emitting layer. Therefore, the self-quenching or triplet–triplet annihilation by dopant molecules is a significant problem in host-guest systems with high doping concentrations. Earlier, Kawamura et al. [40] reported that when the doping concentration is increased from 2 to 6 % the phosphorescence quantum efficiency of $Ir(ppy)_3$ is decreased by ~5 %. Consequently, the selection of suitable host candidates was a critical issue in fabricating high efficiency PHOLEDs.

More recently, to overcome these constraints, red and green PHOLEDs using narrow band gap host materials have been demonstrated, achieving [41, 42]:

1. *a low driving voltage*
2. *high Efficiency (lm/W)*
3. *no charge trapping in phosphorescent guest molecules*
4. *low doping concentration*
5. *low manufacturing cost.*

Several materials having the above characteristics are commercially available. A low phosphorescent guest doping concentration (approximately 1 ~ 2 %) is ideal for harvesting triplet and singlet excitons to produce phosphorescent emission [43, 44].

Summarizing, PHOLEDs have had several problems in the past; particularly, a relatively complex architecture (multilayer structure), relatively high driving voltage, and high roll-off characteristic which are not suitable for AMOLED applications. However, the causes of these issues were discovered and most problems were solved by intensive research. Now good phosphorescent materials and devices suitable for AMOLED applications are commercially available. We can thus forecast that PHOLEDs suitable for high brightness displays and lighting applications will be used commercially in the near future.

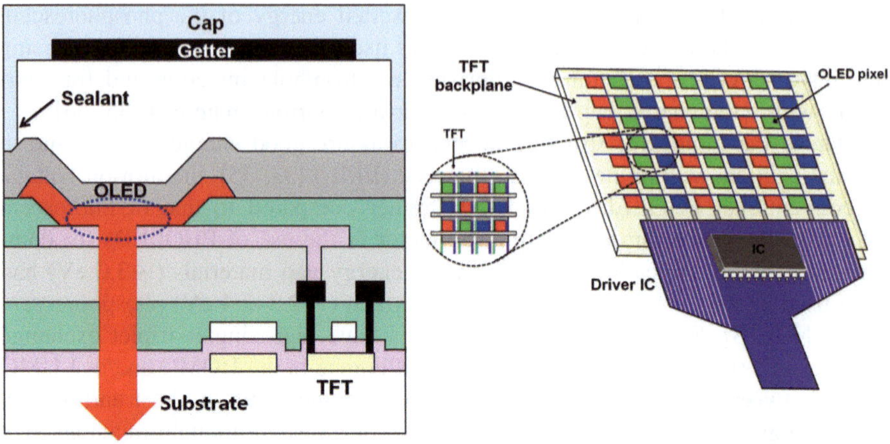

Fig. 3.4 Cross-section and top view of *AMOLED*s. (*TFT* = Thin film transistor)

3.5 Current Status of OLED Technology for AMOLED Displays

Figure 3.4 shows cross-section and top view of an AMOLED structure. A TFT backplane is built on a glass substrate and the pixel electrode connects to the drain of the TFT. The edge of the pixel electrode is covered with an insulator material. The multi-layered red, green and blue OLEDs are connected to the pixel electrodes. From the end of 1990, AMOLEDs have been studied intensively for mobile and TV display applications. Mobile AMOLED displays for devices like smart phones, cellular phones, and personal multimedia players have already been in the market for several years. 11 and 15 inch AMOLED TVs by Sony and LG display have been successfully launched for pre-marketing purposes already in Dec. 2007 and Nov. 2009, respectively, and 55" displays have been shown in 2012. AMOLEDs have outstanding display qualities such as high resolution, very good contrast ratio, low power consumption and good video image quality. The latter arises particularly from the fast response to changing images and good color purity ensured by OLEDs. Furthermore, the AMOLED displays are specially suited to mobile applications since they are extremely thin, operate over a wide temperature range, and have a wide viewing angle. These good performance characteristics are fast gaining to AMOLEDs a strong market position. Current fabrication technologies for manufacturing mobile AMOLEDs typically use a low temperature polysilicon (LTPS) backplane, vacuum sublimation of organic molecules, a shadow mask process for pixel patterning, and a top emission structure.

Research for next generation applications is now focusing on large screen AMOLED display applications. These applications must be manufactured at a cost which enables competition with active matrix liquid crystal display (AMLCD).

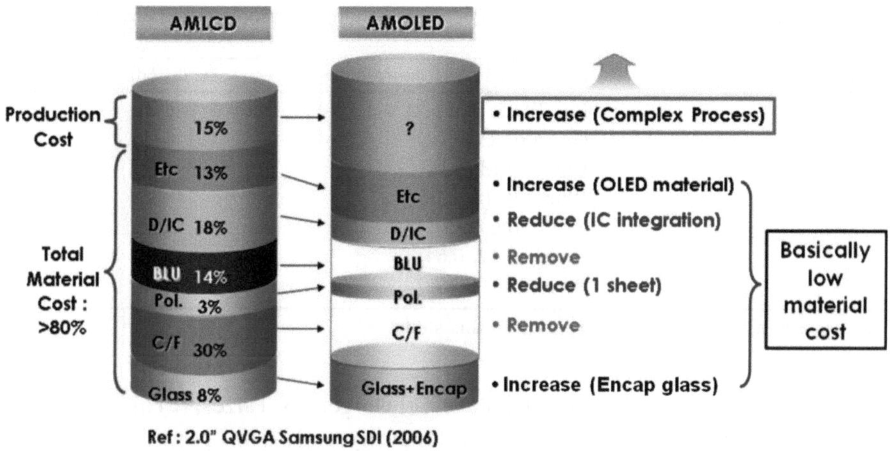

Fig. 3.5 Cost comparison between *AMOLED* and *TFT LCD* (*AMLCD*) displays. In *AMOLED*s, the back light unit and color filter (*C/F*) are not necessary and the cost of driver *IC* (integrated circuit) (*D/IC*) and of the polarizer (*pol.*) can be reduced, but the depreciation and manufacturing costs are high

Figure 3.5 shows a cost comparison between AMOLED and AMLCD for mobile applications. Current commercial OLED technology is using relatively small size glass substrates, so-called "generation 3.5" (730×460 mm^2) and "1/4 of generation 5.5" (750×650 mm^2), together with a photo mask process which is not as simple as for AMLCD. Also, the depreciation and manufacturing costs are high albeit the total number of parts in the display is reduced: in AMOLEDs, high cost parts like the color filter (C/F) and the back light unit are not necessary. In order to have a low cost structure of AMOLEDs, a large area TFT backplane technology with a low mask count and a pixel patterning technology suitable for large size substrates is the subject of active research right now. The current technology status of AMOLED for large screen TV applications is discussed more in detail in the following subsections.

3.5.1 Top Emission Technology

Earlier investigations were focused on bottom emission OLEDs structures where the light is emitted through a transparent bottom electrode (generally ITO) and the glass substrate (Fig. 3.3, left). The major advantage of a bottom emission arrangement is the simplicity of the device structure. However, the emission aperture has to share the substrate with the electronics needed to drive the light-emitting device, thereby potentially limiting the pixel size in case of high resolution displays. This could be avoided by employing top emission structures (Fig. 3.3 right), where the light escapes from the device through a transparent cathode and encapsulation. The top

Table 3.2 Efficiency and color comparison between bottom emission and top emission devices [48]

Emission mode	Color (CIE 1931 coordinates)	Efficiency (cd/A)
Top	Red (0.65, 0.35)	32
	Green (0.23, 0.72)	26.2
	Blue1 (0.13, 0.14)	3.8
	Blue 2 (0.14, 0.06)	1.5
Botttom	Red (0.65, 0.35)	12
	Green (0.32, 0.64)	12
	Blue (0.14, 0.16)	5

The same emissive materials were used in this study

emission leads to the following advantages: (1) larger pixel sizes, (2) high efficiency at the front viewing angle, (3) saturated color purity, and (4) low power consumption.

Several pros and cons of the top emitting structure have been reported in [45, 46]. The optical out-coupling efficiency (i.e. external quantum efficiency) and color purity can be improved significantly due to micro-cavity effects via light multi-reflections between the reflective anode and semi-transparent cathode. The multi-interference technology can improve the color gamut of OLEDs without changing the organic materials. Tokito et al. [45] reported pure red, green and blue emission in OLEDs with a planar micro-cavity effect defined by a dielectric mirror/metal mirror pair. Pure red, green and blue emission with (0.67, 0.33), (0.24, 0.72), and (0.18, 0.05) color coordinates were obtained by controlling the structure of the dielectric mirror. Similar results were also reported by Dodabalapur et al. [46] using a dielectric mirror. Reducing wave-guiding losses using a high refractive index material layer on top of a semi-transparent cathode can also improve substantially the device efficiency. Exploiting micro-cavities together with improved wave-guiding loss the device luminous efficiency could be enhanced by a factor two for green and by a factor three for red emission [47, 48]. Table 3.2 shows efficiency comparison data between bottom and top emission devices [48]. Current AMOLED mobile phones using the top emission device structure have nearly 100 % color gamut against NTSC color coordinates.

The disadvantages of the top emission structure are the manufacturing difficulty, color coordinate variation, and efficiency variation. The optical output obtained by micro-cavity effect varies significantly depending on the total organic layer thickness. The following equation describes micro-cavity effects in relation to optical output. The total optical thickness of the cavity, L is given by

$$L(\lambda) \approx \frac{\lambda}{2}\left(\frac{n}{\Delta n}\right) + \sum_j n_j L_j + \left|\frac{\varphi_m}{4\pi}\lambda\right|$$

The first term in this equation is the penetration depth of the electromagnetic field into the mirror dielectric stack where λ is the center wavelength, n is average refractive index, and Δn is index difference between the layers. The second term is the sum of optical thicknesses of organic layers between the two mirrors. The last term is the effective penetration depth into the top metal mirror, where φ_m is the phase shift at the metal reflector. From this expression, we know that the exact control of the organic layer thickness is important to obtain uniform optical characteristics across a large area substrate. In a real production environment it is challenging to obtain good thickness uniformity on a large substrate. However, precise process control of the organic layer thickness in production has been already established on half-size generation 3.5 (740 × 460 mm^2) and 1/4 of generation 5.5 (750 × 650 mm^2) substrates. The OLED color gamut in large area applications is about 70 % of the NTSC color range [23–25] using currently available materials; further enhancements are required to enable top emission large area TV applications. Recently, micro-cavity effects have also been used with bottom emission OLEDs to achieve 100 % color gamut. Figure 3.6 shows a prototype Samsung 31" AMOLED display which employs micro-cavity effects with bottom emission. Enhanced color technology utilizing micro-cavity effects will soon be employed in AMOLED TV products.

3.5.2 Pixel Patterning Technology

Currently, AMOLED pixels are manufactured using 50 μm thick metal shadow masks. When each color material is deposited, other color areas are blocked by this shadow mask. Figure 3.7 shows schematically the shadow mask technology for red, green, blue pixel patterning of AMOLEDs. This mask is produced primarily using chemical etching, laser etching, or eletro-deposition. Potential mask sagging problems can be overcome using a mask tension process. However, the sub-pixel width obtained with shadow masking can vary by ±10 μm due to etching non-uniformity. This value is exceptionally large compared with other AM display technologies. Other factors contributing to accuracy issues in fabrication are the total pitch variation, shadow effects due to a thick mask, and alignment accuracy between the mask and substrate. Regular mask cleaning prevents additional pixel window variations that may be induced by the continuous deposition process. The printing accuracy, including all possible process errors, is currently about ±15 μm.

Manufacturing of generation 2 (370 × 470 mm^2) till 1/4 of generation 5.5 (750 × 650 mm^2) substrates with the current mask technology is already mature and does not present serious problems. However, this technology cannot be used for generation 7–8 substrates. A new approach to use the shadow mask process for large area TV applications has been developed using a substrate scanning technology where the glass substrate is moved during the pixel deposition while keeping stationary both the shadow masks and the evaporation sources. Samsung has demonstrated at the Consumer Electronics Show (CES) 2012 a 55" AMOLED display using this glass scanning technology—Fig. 3.8. This panel is driven using LTPS TFTs.

31" AMOLED Prototype (Samsung, SID 2008)

Items	Specification
Size	31"
Resolution	Full HD
Panel Thick.	4.8mm
Pixel pitch	117 x 351 mm
Emission	Bottom
Peak Bright	> 400 cd/m^2
Color	16.7 M color
Contrast	>1000:1

Fig. 3.6 31" *AMOLED* display made with micro-cavity technology

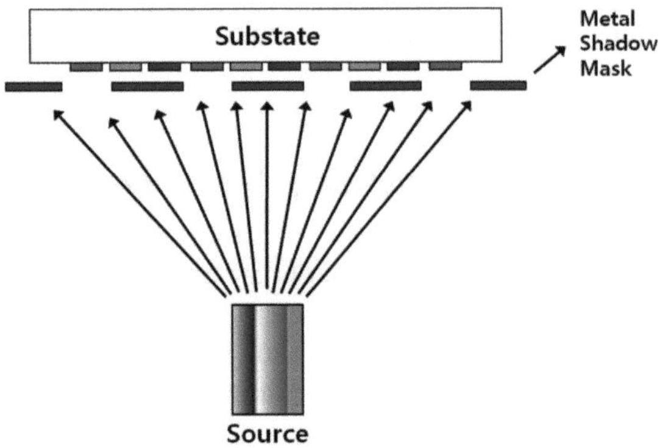

Items	Current Status (generarion 3.5 evaporator)	
Shadow Effect	〈 5 μm	Pixel-Pixel Align Margin : ±15μm
Pixel Pitch Variation	±10 μm	
Total Pitch Variation	±5 μm	
Alignment Accuracy	〈 ±5 μm	

Fig. 3.7 Shadow mask technology for pixel patterning of *AMOLED*s. The table shows the current technical status of the shadow mask process

Several companies are engaged in developing other technologies to replace the current mask process for large area TV applications. Solution process printing technologies such as ink-jet printing [49], nozzle printing [50, 51] and several laser induced printing technologies [52–54] have been also proposed as pixel patterning methods suitable for large areas. Laser induced thermal imaging (LITI) technology is the first reported method using a laser printing approach and is still under development [52]. This technology uses donor films having a laser light absorbing layer and a transfer organic layer. Laser light is converted into thermal energy as it shines on the light absorbing layer. Subsequently, the transfer layer melts and transfers to the substrate. Figure 3.9 shows schematically the laser induced thermal imaging process. Radiation induced sublimation transfer (RIST) technology [53]

Fig. 3.8 A Samsung 55" AMOLED display that was demonstrated at the Consumer Electronics Show 2012

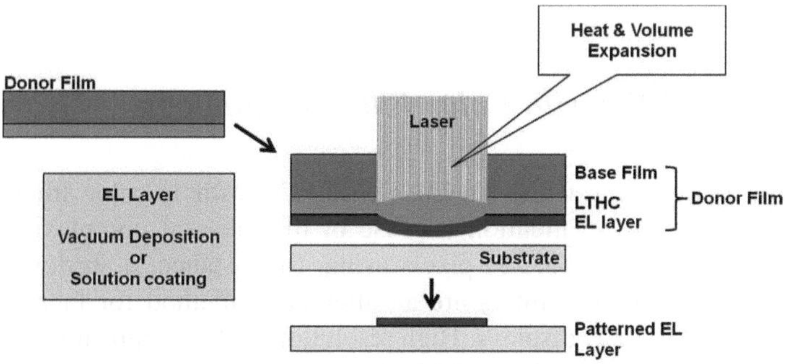

Fig. 3.9 Laser induced thermal imaging process for pixel patterning of *AMOLED*s. (*LTHC* = light to heat conversion layer, *EL* = Layer electroluminescence layer)

and laser induced pattern-wise sublimation (LIPS) technology [54] were also reported as alternatives to LITI technology. Both technologies are basically very similar. Emissive organic layers on the light absorbing metal layer are deposited on glass and then attached to two substrates, an organic deposited substrate and an AM backplane under vacuum. Finally a laser is scanned across the organic layer to form a pattern through sublimation transfer from the organic layer to the AM backplane. Current process issues using these laser technologies are thermal damage, process stability and yield. However, the rapid development of these laser technologies is continuing using large substrates.

The ultimate objective is to fabricate a pixel patterned OLED using solution processing since this approach is simple and does not require any vacuum equipment. Ink-jet printing and nozzle printing technologies are potential candidates for solution processing. Several material companies are developing soluble

Table 3.3 Device characteristics of solution processed OLEDs

Color	CIE 1931 coordinates	Efficiency (cd/A)	Lifetime@1,000nit (h)
Red	(0.63,0.37)	30	>200,000
	(0.66,0.33)	9	100,000
Green	(0.29,0.64)	18	80,000
	(0.33,0.63)	35	63,000
Blue	(0.14,0.18)	12	18,000
	(0.19,0.40)	18	6,000

materials for these solution processes. However, it is not an easy task to develop good soluble materials because OLED lifetime is very sensitive to impurities, film quality, and environmental conditions. The technical status of current soluble materials is summarized in Table 3.3 [50, 55, 56]. These data were obtained by spin coating soluble materials and are not the result of printed pixel performance. Recently, a significant progress has been made in the development of soluble materials for device fabrication. Hence, it is expected that the device grade solution printing processes will be available in the near future.

3.5.3 White OLED with Color Filter Technology

The approach of using white OLED (WOLED) with a color filter (CF) has received a great deal of attention because of the inherent complexity of manufacturing red–green–blue (RGB) pixels using, for instance, a shadow mask process. WOLEDs with color filters are an alternative method for fabricating large area full color AMOLED displays. High resolution mobile applications can also be realized using this simple white OLED deposition process.

White emission can be achieved by mixing three primary colors (red, green, and blue) or two complementary colors from different emitters. Generally, WOLEDs consisting of small organic molecules have a multilayer structure with two or more emitting layers in a stack [57]. This CF approach reduces the light output of about 70 % because of the light absorption caused by CF pigments. This problem can be addressed using the RGBW[3] color system [58–60]. The RGBW system is approximately two times more efficient than the RGB system owing to the 100 % transmittance of the white pixel. The power efficiency is not a real concern in TV applications even if it would be worse than that of existing technologies. On the

[3] RGBW: RGB color filters transmit only a small band of wavelengths, and thus only a small percentage of the light generated by an OLED light is visible to the eye, which reduces brightness. The RGBW pixel system has a white sub-pixel to the RGB mix that is actually a clear area in the OLED display with no color filter material; therefore nearly all of the light is transmitted through the white sub-pixel. This system can achieve higher efficiency and brighter displays.

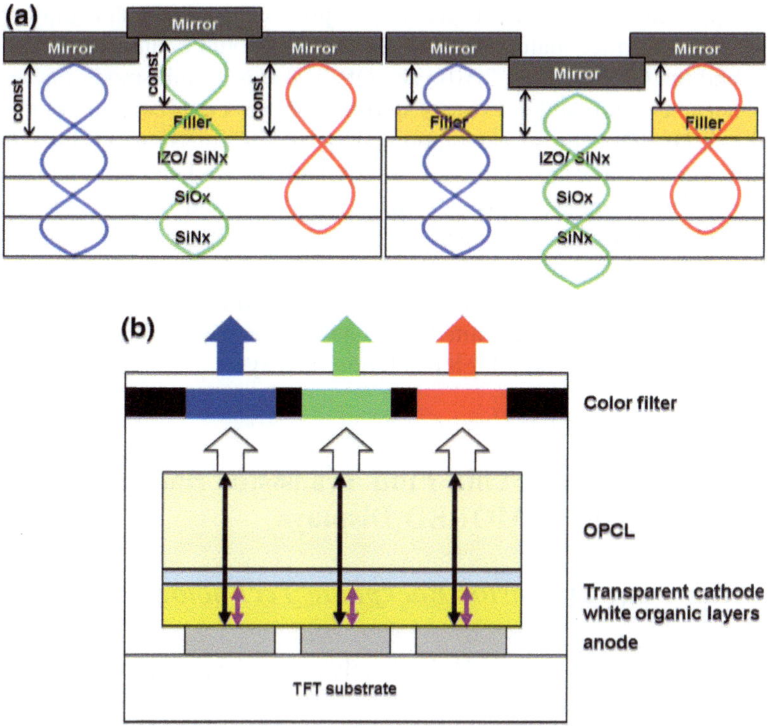

Fig. 3.10 **a** Bottom emission white *OLED* approach with *CF*. **b** Top emission white *OLED* approach with *CF*. (*OPCL* = optical path control layer)

contrary, a wide color gamut is essential for commercialization; a 100 % color gamut using new color filter pigments was developed by Spindler et al. [61], although in this work there were some drawbacks related to the efficiency and lifetime of the RGB subpixels. These issues have been overcome by Lee et. al [62] using a new optical white OLED + CF incorporating micro-cavity resonance in the AMOLED panel. These authors used mirror layers on the substrate to fabricate micro-cavity structures. This design produced a color gamut >100 % and an (R + G + B)/W ratio of 120 % [61]. Figure 3.10 shows schematically white OLED + CF optical designs for AMOLED panels.

The top emission structure approach is most desirable to fabricate ultra high resolution displays since the aperture ratio of the bottom emission OLED is <10 % for a panel with a resolution >300 ppi. With the current shadow mask technology, it is impossible to fabricate a high resolution display since etching non-uniformity, tension variation, and the shadow effects induced by the thick metal mask only allow patterning with a resolution lower than 250 ppi. If a top emission white OLED + CF approach is used it should be possible to produce a panel with a resolution >200 ppi, however, using top emission WOLED it is difficult to get clearly separated RGB peaks. Therefore, to produce a wide color gamut it was proposed that an optical path control layer (OPCL) should be inserted between the cathode and the CF [63] (Fig. 3.10). The OPCL forms a multi-mode micro-cavity in the device structure which extracts separated sharp red, green, and blue spectra from the WOLED. Kim et. al reported [64] a top emission 3.0 inch WVGA AMOLED prototype panel with RGB peaks which realized an excellent resolution of 308 ppi. The panel was driven using LTPS TFTs and had a color gamut of >90 %. LG Display had demonstrated a 55" AMOLED display exploiting the white OLED + CF approach at the Consumer Electronics Show (CES) 2012. This panel was fabricated using the RGBW system and oxide TFTs technology for the backplane. In conclusion, the white OLED with color filter technology has a great potential to enable large area high resolution displays.

3.6 Current Status of Thin Film Transistor Backplane Technology for AMOLED Displays

3.6.1 Thin Film Transistor Backplane Technology

As the luminescence of AMOLEDs is directly proportional to the electrical current flowing through the LED device, which is controlled by the thin film transistors (TFTs), the TFT backplanes for AMOLEDs require uniform device performance to prevent brightness variations within the display and enable reliable device performance, sustaining a constant current flow even after a long time of operation. Three different types of TFT backplanes for AMOLEDs have been considered till now: a-Si:H TFTs [65], low temperature polycrystalline silicon (LTPS) TFTs [66], and oxide TFTs [67, 68].

Although a-Si:H TFTs are scalable and can be fabricated with a high degree of uniformity for LCD applications, their device performance, particularly threshold voltage, is seriously degraded under the constant current stress required for AMOLEDs operation. The threshold voltage shift of a-Si:H TFT with respect to time results in the 'image sticking' problem in AMOLEDs: a change in the amount of current supplied by transistor under the same biasing conditions. Because the amount of threshold voltage shift depends on the level of constant current stress of each pixel, the previous image sent to the display remains after refreshing the frame. It looks thus as if the 'image sticks' to the AMOLED display, an effect that seriously degrades the display image.

Low temperature polycrystalline silicon (LTPS) is also a proven technology used to manufacture AMLCDs panels. LTPS TFTs fabricated by excimer laser annealing (ELA) is a backplane technology for AMOLEDs which is commercially available and is considered to be the most suitable technology for producing high stability devices. This backplane technology is currently used in mobile applications. However, the ELA based LTPS TFT technology have limitations in terms of scalability and uniformity when used for large area applications like TVs. Even though recently the maximum beam length is increased from 460 to 750 mm, this process is not very cost-effective for TV applications. Furthermore, fluctuations in laser processing cause variations of the TFT performance and result in image non-uniformity in AMOLEDs. In order to improve image non-uniformity, several circuit techniques compensating the variation of TFT properties have been developed, as will be discussed in Sect. 3.6.2.

Other methods of fabricating LTPS TFTs have been investigated over several years. Some examples include fabrication using thermal annealing (solid phase crystallization) and metal induced crystallization; they have fairly good scalability and long range uniformity. However, these technologies also have other drawbacks. Thermal annealing processes with temperatures >650 °C can damage the glass substrates and the poor silicon crystallinity of laser-free crystallization processes results in high TFT leakage currents. Therefore, it is important to optimize the process temperature in order to minimize the glass deformation and to suppress the leakage current in the TFTs.

Recently, amorphous semiconductor-oxides have attracted huge attention as promising materials to replace conventional silicon in TFTs. The amorphous nature of oxide TFTs produces devices with good uniformity. Oxide TFTs have also superior performance in terms of device mobility and stability in comparison to a-Si:H TFTs. Moreover, the architectures and process equipment used to fabricate oxide TFTs are similar to those used in the fabrication of a-Si:H TFTs; thus processing is scalable and compatible. In order to exploit oxide TFTs for commercial production of AMOLED TVs it is necessary to improve the stability of these devices under DC bias, light and temperature stressing. Furthermore, considering the manufacturability of large area AMOLEDs at an affordable price, the size of TFT substrates should be comparable with those of LCDs. The number of panels fabricated on a substrate is key factor to reduce the fabrication cost of TFT, as it shares the depreciation cost of the huge investments needed for the TFT production lines.

Recently Samsung has shown a 55"AMOLED (Fig. 3.8) exploiting LTPS TFT technology. Samsung will thus probably produce AMOLED TVs with LTPS TFTs in the near future. On the other hand, still at CES2012, LG Display has demonstrated a 55" AMOLED with oxide TFT technology. A strategic choice about the TFT technology which is most suited for mass production is thus still ongoing, and it may be that different companies will follow different development roadmaps.

3.6.2 OLED Driving Technology

OLEDs are current driven devices whereas LCDs are voltage driven. In the latter liquid crystals are modulated by an applied electric field and act as shutters for the light which comes from a back light unit. Both the passive and active matrix driving schemes used in LCDs can also be used in OLEDs. The passive matrix

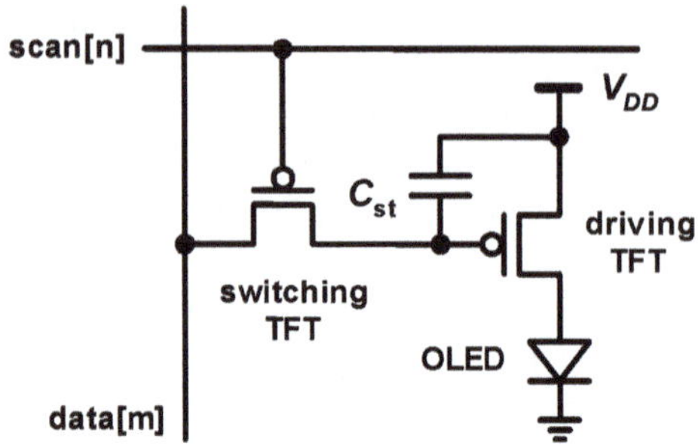

Fig. 3.11 Conventional pixel circuit for *AMOLED*, consisting of two *TFT*s and one capacitor

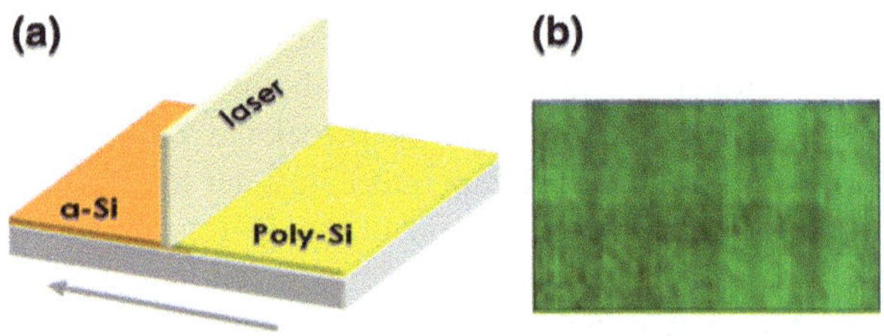

Fig. 3.12 **a** A conceptual drawing describing crystallization procedure of silicon from amorphous to poly crystalline micro-structure by ELA process. **b** Brightness non-uniformity originated from the variations in the ELA process

OLEDs (PMOLEDs) are low cost devices because they consist of simple X–Y matrix of electrodes. A PMOLED pixel is illuminated when it is selected by X–Y electrode signals. However, the PMOLEDs have significant disadvantages. Since the PMOLEDs utilize transparent electrodes with high resistivity, and large parasitic capacitances are associated with the crossing of X–Y electrodes, they inevitably consume too much electric power and cannot realize good image uniformity for large area TVs with high resolution.

On the other hand, AMOLEDs which incorporate TFTs controlling each pixel individually have a wider range of applications for high resolution, small to large area displays. In order to supply power to the AMOLED display, additional power-supply lines and power-control transistors are required for each pixel. Therefore, for each AMOLED pixel at least two transistors and one capacitor are required (Fig. 3.11). The switching TFT enables the pixel under control of the input signals and the driving TFT modulates the power supplied to the OLEDs when it is enabled. The brightness of AMOLED devices increases linearly with an increase in the current supplied from the driving TFT. Hence, even a slight non-uniformity in TFT performance can create serious image quality problems (Fig. 3.12). In order to correct this TFT-related non-uniformity, compensation circuits are generally adopted for each pixel.

Two types of compensation circuits have been suggested; current programming [69, 70] and voltage programming [71]. Figure 3.13 shows a circuit diagram and timing charts for a representative current programming circuit. During the current programming period, the gate voltage of the driving TFT (T1) is set by a data current I_{data} that flows in T1 while it is diode-connected, and stored on the

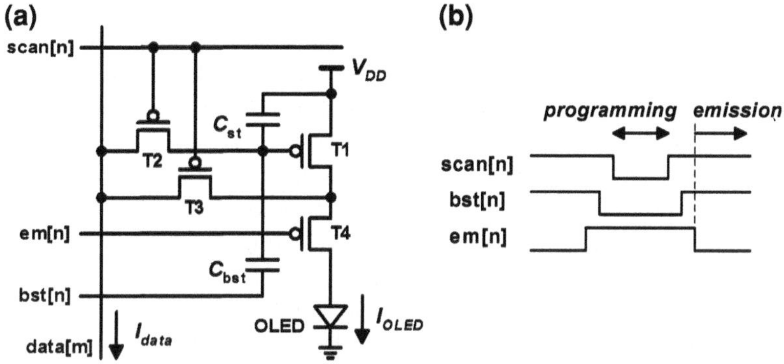

Fig. 3.13 **a** A current programming compensation circuit configuration disclosed by Samsung Co. **b** Driving waveforms of the circuit

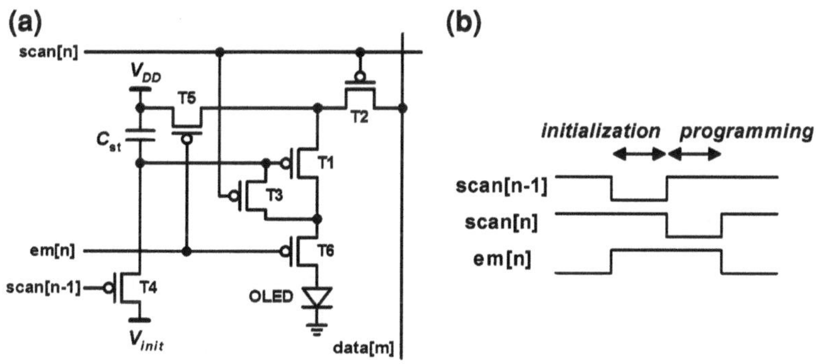

Fig. 3.14 **a** A voltage programming compensation circuit configuration disclosed by Samsung Co. and Hanyang University. **b** Driving waveforms of the circuit

capacitor C_{st}. If the bootstrap input voltage Boost is not used, in the emission phase the current flowing through T1 exactly matches the programming current, compensating any variation in both threshold voltage and mobility of T1. When the capacitive bootstrap bst is used, the current flowing into the OLED during the emission time can be decreased compared to the programming current, but the mobility cancellation is only partial (see [70]). Figure 3.14 shows a representative example of a voltage programming compensation circuit. During the programming period, a voltage $V_{data} - |V_{th}|(T1)$ is stored at the gate node of the driving TFT (T1) switching on the TFT T3. In the emission phase, $V_{sg}(T1)$ becomes $(V_{dd} - V_{data} + |V_{th}|(T1))$ and, since the current of T1 is proportional to $(V_{sg} - |V_{th}|(T1))^2$, the emission current of T1 will be proportional to $(V_{dd} - V_{data})^2$. In this way the non-uniformity of threshold voltage of the driving TFTs can be effectively compensated.

The current programming circuit of Fig. 3.13 uses the data current and, in general, the large load capacitance of the data line must be charged during the programming period. It is thus difficult to compensate the deviations of the driving TFTs at low gray levels, at which only a small amount of current should flows through the OLED. Therefore, the current programming method has to be further developed to be adapted to large area AMOLED TVs. On the contrary, although voltage programming compensation circuits can compensate only the variation in the threshold voltage of the driving TFTs, most AMOLED products on the market utilize voltage programming compensation circuits because this scheme can be implemented with minor modifications of the existing LCD driver ICs.

Finally, digital compensation circuits have been suggested by several authors [72–74]. Digital circuits use transistors either with a large gate overdrive or below threshold, and are thus less sensitive to threshold variations.

The gray level of the AMOLEDs with digital driving schemes can be adjusted by controlling the emission time, using so called time-division methods. However, since the OLED materials have to always emit at full brightness in this case, the lifetime of OLED materials has to be improved further to apply with success digital driving circuits to large area and high resolution AMOLEDs.

3.7 Future Directions

After several years of AMOLED display mass production (since 2007), AMOLED production yield has reached a level comparable with LCDs. The possible applications range from small area (cellular phones) to medium area (navigation systems). AMOLED displays have proven superior performance in mobile applications due to their vivid image quality, excellent viewing angle, fast response time and so on. Nevertheless, the production capacity has to be significantly enhanced to attain cost competitiveness with LCD.

The most promising application of AMOLEDs, however, is large area TVs. Several companies have already started AMOLED TV production and demonstrated the beauty of AMOLED TVs. In addition to their outstanding performance, AMOLEDs are the most eco-friendly display because they consume much less material, reducing or eliminating many parts used in LCDs like polarizer, backlight unit and color filter. Moreover, their power consumption can be lowered by turning on only the selected pixels of the display whereas AMLCD uses the backlight on the whole area of the display (or in selected areas when the local dimming technology is utilized).

The realization of AMOLED TVs, however, still faces several challenges that come from scaling-up the size of glass and reducing process cost. Moreover, OLED performance still requires some improvements for TV applications. From the backplane point of view, the available technologies such as a-Si and ELA-based LTPS TFTs have restrictions in terms of device stability and scalability, respectively. Recently, metal-oxide TFTs have been in the spotlight as a possible replacement for Si-based TFTs for large area AMOLED displays. Replacement with oxide TFTs is considered not only because they have the required device

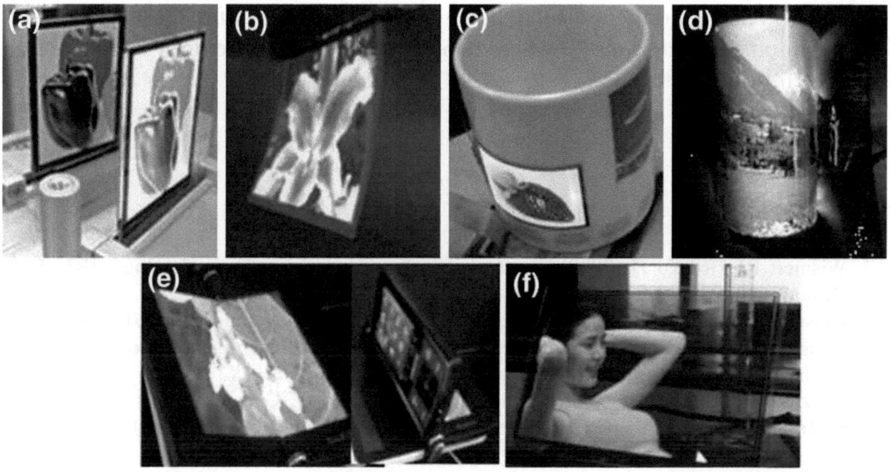

Fig. 3.15 Various unique applications of *AMOLED* demonstrated by Samsung Co. **a** Emission on both sides (SID (The Society for Information Display) 2004 exhibition). **b** Flapping (FPD international 2008 exhibition). **c** Bendable (SID 2007 exhibition). **d** Flexible (FPD international 2008 exhibition). **e** Foldable (SID 2008 exhibition). **f** Transparent (SID 2008 exhibition)

stability and scalability, but also because the conventional processes and process equipment used for LCD fabrication can be used without additional supplements with these materials.

From the OLED point of view, the commercially available shadow mask technology has scalability and material utilization efficiency issues for large area AMOLEDs. Therefore, there has been a strong demand for ink-jet or nozzle printing of OLEDs to enable high scalability and material utilization efficiency. However, the current performance of solution-processed materials required for printing is insufficient to be used for large area AMOLED applications. Therefore, the search of innovative solution-processed materials for device fabrication is an attractive research theme.

OLEDs are the first organic devices that have successfully reached the mass production stage. The success of AMOLEDs will spur the development of other organic devices such as organic solar cells, organic sensors and organic TFTs. The OLED mass production technology can indeed easily be transformed for the production of other types of organic devices.

AMOLEDs functionality is achieved with a single substrate, without any need for additional components such as liquid crystal: for this reason AMOLEDS have unique properties that LCDs do not possess. Indeed, AMOLEDs have realized many dream applications that seem to come out from science-fiction movies. Figure 3.15 shows real examples of transparent, bendable, foldable and flexible AMOLEDs. AMOLEDs are likely to replace gradually conventional LCD displays due to their superior properties. But more importantly, they have a special ability to inspire completely novel applications not yet dreamt of. The potential of

AMOLEDs lies not only in display innovation: in the near future these devices will evolve into "*real interfaces*", where people and electronic devices will be able to connect in intuitive ways.

References

1. Wisnieff RL, Ritsko JJ (2000) Electronic displays for information technology. IBM J Res Develop 44:409
2. Krasnov AN (2003) Electroluminescent displays: history and lessons learned. Displays 24:73
3. Hung LS, Chen CH (2002) Recent progress of molecular organic electroluminescent materials and devices. Mater Sci Eng. R Rep 39:143
4. Ortiz S Jr (2003) New monitor technologies are on display. Comput 36(2):13
5. Helfrich W, Schneider WG (1965) Recombination radiation in anthracene crystals. Phys Rev Lett 14:229
6. Tang CW (1982) Organic electroluminescent cell. US Patent No. 4356429 (26th Oct 1982)
7. Tang CW, VanSlyke SA (1987) Organic electroluminescent diodes. Appl Phys Lett 51:913
8. VanSlyke SA, Tang CW (1985) Organic electroluminescent devices having improved power conversion efficiencies. US Patent 4539507
9. Tang CW, Chen CH, Goswami R (1988) Electroluminescent device with modified thin film luminescent zone. US Patent No. 4769292
10. VanSlyke SA, Chen CH, Tang CW (1996) Organic electroluminescent devices with improved stability. Appl Phys Lett 69:2160
11. Shirota Y, Kuwabara Y, Inada H, Wakimoto T, Nakada H, Yonemoto Y, Kawami S, Imai K (1994) Multilayered organic electroluminescent device using a novel starburst molecule, 4,4′,4″-tris(3-methylphenylphenylamino)triphenylamine, as a hole transport material. Appl Phys Lett 65: 807
12. Deng ZB, Ding XM, Lee ST, Gambling WA (1999) Enhanced brightness and efficiency in organic electroluminescent devices using SiO_2 buffer layers. Appl Phys Lett 74: 2227
13. Adamovich VI, Cordero SR, Djurovich PI, Tamayo A, Thompson ME, D'Andrade BW, Forrest SR (2003) New charge-carrier blocking materials for high efficiency OLEDs. Org Electron 4(2–3):77
14. Hung LS, Tang CW, Mason MG (1997) Enhanced electron injection in organic electroluminescence devices using an Al/LiF electrode. Appl Phys Lett 70:151
15. Kido J, Lizumi Y (1998) Fabrication of highly efficient organic electroluminescent devices. Appl Phys Lett 73: 2721
16. Kido J, Matsumoto T (1998) Bright organic electroluminescent devices having a metal-doped electron-injecting layer. App Phys Lett 73: 2866
17. Endo J, Matsumoto T, Kido J (2002) Organic electroluminescent devices with a vacuum-deposited Lewis-acid-doped hole-injecting layer. Jpn J App Phys 41: L358
18. Sato Y, Ogata T, Kido J (2000) Organic electroluminescent devices with polymer buffer layer. Proc SPIE 4105: 134
19. Zhou X, Pfeiffer M, Blochwitz J, Werner A, Nollau A, Fritz T, Leo K (2001) Very-low-operating-voltage organic light-emitting diodes using a p-doped amorphous hole injection layer. App Phys Lett 78:410
20. Huang J, Pfeiffer M, Werner A, Blochwitz J, Leo K, Liu S (2002) Low-voltage organic electroluminescent devices using pin structures. Appl Phys Lett 80:139
21. He G, Scvhneider O, Qin D, Zhou X, Pfeiffer M, Leo K (2004) Very high-efficiency and low voltage phosphorescent organic light-emitting diodes based on a p-i-n junction. J App Phys 95:5773

22. Wellermann P, Hofmann M, Zeika O, Werner A, Bimstock J, Meerheim R, Walzer GF, He K, Pfeiffer M, Leo K (2005) High-efficiency p-i-n organic light-emitting diodes with long lifetime. J Soc Inf Disp 1:393
23. Arakane T, Funahashi M, Kuma H, Fukuoka K, Ikeda K, Yamamoto H, Moriwaki F, Hosokawa C (2006) Fluorescent RGB OLEDs with high performance. SID 2006 Digest 37
24. Kawamura M, Kawamura Y, Mizuki Y, Funahashi M, Kuma H, Hosokawa C (2010) Highly efficient fluorescent blue OLEDs with efficiency-enhancement layer. SID 2010 Digest 560
25. D'Andrade BW, Weaver MS, Mackenzie PB, Yamamoto H, Brown JJ, Giebink NC, Forrest SR, Thompson ME (2008) Blue phosphorescent organic light emitting device stability analysis. SID 08 Digest, p 712
26. Aziz H, Popovic Z, Tripp CP, Hu N-X, Hor A-M, Xu G (1998) Degradation processes at the cathode/organic interface in organic light emitting devices with Mg:Ag cathodes. Appl Phys Lett 72:2642
27. Nishimura K, Kawamura Y, Kato T, Numata M, Kawamura M, Ogiwara T, Yamamoto H, Iwakuma T, Jinde Y, Hosokawa C (2009) New green and red phosphorescent host materials for highly-efficient and long-lifetime OLEDs. SID 2009 Digest, p 310
28. Tsuzuki T, Shirasawa N, Suzuki T, Tokito S (2003) Color tunable organic light-emitting diodes using pentafluorophenyl-substituted iridium complexes. Adv Mater 15:1455
29. Adachi C, Baldo MA, Forrest SR, Thompson ME (2000) High-efficiency organic electrophosphorescent devices with tris(2-phenylpyridine)iridium doped into electron-transporting materials. App Phys Lett 77: 904
30. Sun Y, Giebink NC, Kanno H, Ma B, Thompson ME, Forrest SR (2006) Management of singlet and triplet excitons for efficient white organic light-emitting devices. Nature 440:908
31. Baldo MA, O'Brien DF, You Y, Shoustikov A, Sibley S, Thompson ME, Forrest SR (1998) Highly efficient phosphorescent emission from organic electroluminescent devices. Nature (London) 395:151
32. Adachi C, Baldo MA, Forrest SR, Lamansky S, Thompson ME, Kwong RC (2001) High-efficiency red electrophosphorescence devices. Appl Phys Lett 78:1622
33. Baldo MA, Lamansky S, Burrows PE, Thompson ME, Forrest SR (1999) Very high-efficiency green organic light-emitting devices based on electrophosphorescence. App Phys Lett 75:4
34. Adachi C, Baldo MA, Thompson ME, Forrest SR (2001) Nearly 100% internal phosphorescence efficiency in an organic light-emitting device. J App Phys 90:5045
35. Adachi C, Kwong RC, Djurovich P, Adamovich V, Baldo MA, Thompson ME, Forrest SR (2001) Endothermic energy transfer: a mechanism for generating very efficient high-energy phosphorescent emission in organic materials. App Phys Lett 79:2082
36. Holmes RJ, Forrest SR, Tung Y-J, Kwong RC, Brown JJ, Garon S, Thompson ME (2003) Blue organic electrophosphorescence using exothermic host–guest energy transfer. App Phys Lett 82:2422
37. Chin BD, Suh MC, Kim MH, Lee ST, Kim HD, Chung HK (2005) Carrier trapping and efficient recombination of electrophosphorescent device with stepwise doping profile. Appl Phys Lett 86:133505–133507
38. Tsuzuki T, Tokito S (2007) Highly efficient and low-voltage phosphorescent organic light-emitting diodes using an iridium complex as the host material. Adv Mater 19:276–280
39. Gong X, Ostrowski JC, Moses D, Bazan GC, Heeger AJ (2003) Electrophosphorescence from a polymer guest–host system with an Iridium complex as guest: Förster energy transfer and charge trapping. Adv Funct Mater 13:439–444
40. Kawamura Y, Goushi K, Brooks J, Brown J, Sasabe H, Adachi C (2005) 100% phosphorescence quantum efficiency of Ir(III) complexes in organic semiconductor films. Appl Phys Lett 86:071104
41. Park TJ, Jeon WS, Park JJ, Kim SY, Lee YK, Jang J, Kwon JH, Pode R (2008) Efficient simple structure red phosphorescent organic light emitting devices with narrow band-gap fluorescent host. Appl Phys Lett 92:113308–113310
42. Jeon WS, Park TJ, Park JJ, Kim SY, Jang J, Kwon JH, Pode R (2008) Highly efficient bilayer green phosphorescent organic light emitting devices. Appl Phys Lett 92:113311–113313

43. Jeon WS, Park TJ, Kim SY, Pode R, Jang J, Kwon JH (2009) Ideal host and guest system in phosphorescent OLEDs. Org Elect 10:240
44. Kim H-K, Byun Y-H, Das RR, Choi B-K, Ahn P-S (2007) Small molecule based and solution processed highly efficient red electrophosphorescent organic light emitting devices. Appl Phys Lett 91:093512
45. Tokito S, Tsutsui T, Taga Y (1999) Microcavity organic light-emitting diodes for strongly directed pure red, green, and blue emissions. Appl Phys Lett 86:2407
46. Dodabalapur A, Rothberg LJ, Jordan RH, Miller TM, Slusher RE, Philips JM (1996) Physics and applications of organic microcavity light emitting diodes. J Appl Phys 80:6954
47. Lin C-L, Lin H-W, Wu C-C (2005) Examining microcavity organic light-emitting devices having two metal mirrors. Appl Phys Lett 87:021101
48. Hsu S-F, Hwang S-W, Chen CH, Lee S-H, Lee C-C (2006) Highly efficient top-emitting organic light-emitting devices. SID 06 Digest, 1201
49. Lee D, Chung J, Rhee J, Wang J, Hong S, Choi B, Cha S, Kim N, Chung K, Gregory H, Lyon P, Creighton C, Carter J, Hatcher M, Bassett O, Richardson M, Jerram P (2005) Ink jet printed full color polymer LED displays. SID 05 Digest, 527
50. Feehery WF (2007) Solution Processing of Small-Molecule OLEDs. SID 07 Digest, p 1834
51. O'Regan M (2008) Reducing AMOLED manufacturing costs. IMID/IDMC/ASIA Display 08 Digest, p 27
52. Lee ST, Lee JY, Kim MH, Suh MC, Kang TM, Choi YJ, Park JY, Kwon JH, Chung HK, Baetzold J, Bellmann E, Savvateev V, Wolk M, Webster S (2004) A novel patterning method for full-color organic light-emitting devices: laser induced thermal imaging (LITI). SID 04 Digest, p 1008
53. Boroson M, Tutt L, Nguyen K, Preuss D, Culver M, Phelan G (2005) Non-contact OLED color patterning by radiation-induced sublimation transfer (RIST). SID 05 Digest, p 972
54. Hirano T, Matsuo K, Kohinata K, Hanawa K, Matsumi T, Matsuda E, Matsuura R, Ishibashi T, Yoshida A, Sasaoka T (2007) Novel laser transfer technology for manufacturing large sized OLED displays. SID 07 Digest, p 1592
55. Xia S, Cheon K-O, Brooks JJ, Rothman M, Ngo T, Hett P, Kwong RC, Inbasekaran M, Brown JJ, Sonoyama T, Ito M, Seki S, Miyashita S (2008) Printable phosphorescent organic light emitting devices. SID 08 Digest, p 295
56. Lee J (2008) Technical challenges for polymer OLED display manufacturing. IMID/IDMC/ASIA Display 08 Digest, p 1163
57. Murano S, Kucur E, He G, Blochwitz-Nimoth J, Hatwar TK, Spindler J, Slyke SV (2009) White fluorescent PIN OLED with high efficiency and lifetime for display applications. SID 09 Digest, p 417
58. Lee B, Park C, Kim S, Kim T, Yang Y, Oh J, Choi J, Hong M, Sakong D, Chung K, Lee S, Kim C (2003) TFT-LCD with RGBW color system. SID 03 DIGEST, 1212
59. Lee B, Song K, Yang Y, Park C, Oh J, Chai C, Choi J, Roh N, Hong M, Chung K, Lee S, Kim C (2004) Implementation of RGBW color system in TFT-LCDs. SID 04 Digest, p 111
60. Spindler JP, Hatwar TK, Miller ME, Arnold AD, Murdoch MJ, Kane PJ, Ludwicki JE, Alessi PJ, Van Slyke SA (2006) System considerations for RGBW OLED displays. J SID 14:37
61. Spindler JP, Hatwar TK (2007) Development of tandem white architecture for large-sized AMOLED displays with wide color gamut. SID 07 D Digest, p 89
62. Lee B-W, Hwang YI, Lee H-Y, Kim CW, Ju Y-G (2008) Micro-cavity design of RGBW AMOLED for 100% color gamut. SID 08 Digest, p 1050
63. Lee S, Kim M-G, Song J-B, Kim S-Y, Tamura S, Kang S-K, Kim JM, Choi J, Ha J, Lee S, Chu C, Cho S-W, Cho J-Y, Suh M-C (2008) Highly efficient and wide color gamut white OLED architecture for display application. SID 08 Digest, p 826
64. Kim SY, Kim M-G, Lee SH, Song JB, Tamura S, Kang SK, Kim JM, Cho SW, Cho JY, Suh MC, Kim HD (2008) 3.0-in. 308-ppi WVGA AMOLED by top-emitting white OLED with color filter. SID 08 Digest, p 937
65. Tsujimura T et al (2003) A 20-inch OLED display driven by super-amorphous-silicon technology. SID 03 Digest, pp 6–9

66. Chung HK, Lee KY (2005) Alternative approach to large size AMOLED HDTV. SID 05 Digest, pp 956–959
67. Jeong JK, Jeong JH, Choi JH, Im JS, Kim SH, Yang HW, Kang KN, Kim KS, Ahn TK, Chung H-J, Kim M, Gu BS, Park J-S, Mo Y-G, Kim HD, Chung HK (2008) 12.1-inch WXGA AMOLED display driven by indium-gallium-zinc oxide TFTs array. SID '08 Digest, pp 1–4
68. Lee HN, Kyung JW, Kang SK, Kim DY, Sung MC, Kim SJ, Kim CN, Kim HG, Kim ST (2006) Current status of, challenges to, and perspective view of AM-OLEDs. Proc IDW. Otsu, Japan, pp 663–666
69. Matsueda Y, Shin DY, Kim KN, Ryu DH, Chung BY, Kim HK, Chung HK, Kwon OK (2004) 2.2-in. QVGA AMOLED with current de-multiplexer TFT circuits. IDW '04, pp 263–266
70. Shin DY, Matsueda Y, Chung HK (2005) New current demultiplexer TFT circuits for AMOLED. IEICE Trans Electron, E88-C(11)
71. Komiya N, Oh CY, Eom KM, Jeong JT, Chung HK, Choi SM, Kwon OK (2003) Comparison of Vth compensation ability among voltage programming circuits for AMOLED panels. IDW '03, pp 275–278
72. Inukai K, Kimura H, Mizukami M, Maruyama J, Murakami S, Koyama J, Konuma T, Yamazaki S (2000) 4.0-in TFT-OLED display and a novel digital driving method. SID '00 Digest, pp 924–927
73. Tanada Y, Osame M , Fukumoto R, Saito K, Sakata J, Yamazaki S (2004) A 4.3-in. VGA (188ppi) AMOLED display with a new driving method. SID '04 Digest, pp 1398–1401
74. Tagawa A, Numao T, Ohba T (2004) A novel digital-gray-scale driving method with a multiple addressing sequence for AM-OLED displays. IDW '04, pp 279–282

Chapter 4
High Efficiency OLEDs for Lighting Applications

Completing the Solid State Lighting Portfolio

Reinder Coehoorn, Volker van Elsbergen and Coen Verschuren

Abstract Organic Light-Emitting Diode (OLED) technology is developing as a promising option for large area lighting applications, with basic properties such as efficiency, color stability and lifetime which approach or even exceed those of conventional lighting and inorganic LED technology and with various interesting additional complementing features. In this Chapter, an introduction is given on the development of OLED technology for lighting applications. We discuss the working principles of efficient white multilayer OLEDs, the factors which determine the efficiency, several key elements of the fabrication technology including encapsulation methods, and the state-of-the-art as realized in various institutes and companies.

Keywords OLED · Lighting · Charg-carrier mobility · Efficiency · Lifetime · Fluorescence · Phosphorescence · Fabrication · Encapsulation

4.1 Introduction

Organic Light-Emitting Diodes (OLEDs) are thin-film solid state lighting devices, consisting of an organic semiconducting layer or layer stack which is sandwiched in between two electrode layers. When a voltage is applied across the organic layer,

R. Coehoorn (✉) · C. Verschuren
Philips Research Laboratories, High Tech Campus 4,
5656 AE Eindhoven, The Netherlands
e-mail: reinder.coehoorn@philips.com

V. van Elsbergen
Philips Research Laboratories, Weisshausstrasse 2,
D-52066 Aachen, Germany

E. Cantatore (ed.), *Applications of Organic and Printed Electronics*,
Integrated Circuits and Systems, DOI: 10.1007/978-1-4614-3160-2_4,

Fig. 4.1 OLEDs (warm and cold white) as compared to conventional light sources (compact fluorescent (CFL), incandescent and fluorescent (TL) lamps)

the injection of electrons and holes from the cathode and anode, respectively, gives rise to an electrical current. The subsequent radiative recombination of the injected electrons and holes in the organic layer leads to light emission. Emission in a specific organic compound is usually confined to a relatively broad wavelength range, typically 70–100 nm wide. Using a combination of three organic layers, with predominant blue, green and red emission, white light emission can be obtained. Figure 4.1 shows warm and cold white OLEDs, and a comparison with conventional light sources.

In Fig. 4.2 a more quantitative comparison of OLEDs with other light sources is made. Recent advances have led to values of the luminous efficacy of more than 60 lm W^{-1} at present, using rough foils for improving the light-outcoupling efficiency, or to values well above 60 lm W^{-1} using more special methods for improving the light-outcoupling efficiency, as discussed in Sect. 4.2. The highest possible luminous efficacy for a high quality white light source is approximately 300 lm W^{-1}. The system efficiency given in Fig. 4.2 is taken to be the luminous efficacy expressed as a percentage of this maximum efficacy. Another quantity of interest is the lamp efficiency. OLEDs are by nature large-area light sources, which operate typically at 1000 cd m^{-2} for decorative applications, and at 3000–5000 cd m^{-2} for general lighting applications. No special luminaire for light diffusion and for cooling is then necessary, so that there is no related reduction of the lamp efficiency. Natural cooling at the ambient air is sufficient. White OLED lifetimes can exceed more than 10,000 h, as discussed in more detail in Sect. 4.4.

The relatively broad emission spectra from organic materials make it possible that white OLEDs containing three emitting layers can have a very high Color Rendering Index (CRI), up to 90 or more. The CRI expresses the degree to which emission colors of all wavelengths in the visible range are present, which determines the naturalness and vividness of the colors reflected from illuminated objects, i.e. the "pleasantness" of the light source. It is a number in between 0 (bad) and 100 (excellent).

Whereas to-date standard OLEDs are produced by deposition on a thin rigid glass substrate, the small thickness of the active layers and the electrode layers (only a few hundreds of nanometres in total) makes it possible to produce OLEDs

Technology		Luminous efficacy (lm/W)	System efficiency (%)
Incandescent		12	4
Halogen		24	8
Compact fluorescent		54	18
Fluorescent tube	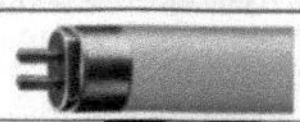	96	32
Inorganic LED's		70-120 (150)	23-40 (50)
Organic LED's		64 (150)	21 (50)

Fig. 4.2 Comparison of various technologies for white lighting applications. For LED and OLED technology, which are still under development, the numbers in between parentheses give the estimated technologically achieveable maximum. For both technologies, the luminous efficacy at brightness levels well below the reference value of 1000 cd m^{-2} can be significantly larger than the values given

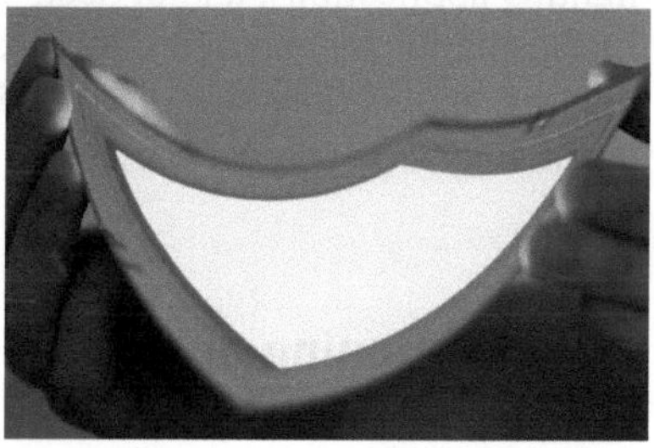

Fig. 4.3 Flexible white OLED (*Source* http://www.holstcentre.com/NewsPress). Photograph reproduced with permission of the Holst Centre (Eindhoven, The Netherlands)

which are flexible and even bendable by deposition on e.g. a polymer foil. Figure 4.3 shows an example of such an OLED. This is expected to open up entirely new areas of applications. Additional demonstrated features of white OLEDs are the operation at safe low voltages (close to the thermodynamic limit of approximately 3 V), easy dimmability, the possibility to tune the emission color, and the possibility to create OLEDs which are transparent in the off-state and emissive to both sides in the on-state (using two conductive transparent electrode layers). At present, color-neutral transparencies of approximately 75 % have been demonstrated.

In view of these prospects, various companies based in the Europe, Asia and the US are developing OLEDs for white lighting applications. A review has recently been presented by Loebl et al. [1]. Philips was, in 2008, the first company to commercialize OLED lighting products. Under the name LUMIBLADE^{TM} (www.the-new-art-of-light.com/), a catalogue collection of OLEDs with different shapes, sizes and colors became available (Fig. 4.4).

Fig. 4.4 OLEDs from Philips Lumiblade, www.the-new-art-of-light.com/

In Sect. 4.2, we discuss the structure and functioning of white OLEDs for lighting applications, and in Sect. 4.3 we discuss the fabrication technologies used. Encapsulation technologies are discussed in Sect. 4.4. In Sect. 4.5 the prospects for lighting applications are discussed. An overview of excellent reviews of OLED technology is given in Refs. [2–8].

4.2 Functioning of OLEDs

OLEDs are based on organic materials within which the bond between the carbon atoms is alternatingly single and double (π-conjugated materials), so that they can function as (intrinsic) semiconductors which conduct when charge is injected from the electrodes. The standard text which describes the physics of electronic processes in organic materials is the book of Pope and Svenberg [9]. Figure 4.5 shows various examples of conjugated organic materials used in OLEDs. As demonstrated first by Tang van VanSlyke [10], it is possible to fabricate OLEDs on the basis of one or more evaporation deposited layers of small molecules, such as aluminium tris(8-hydroxyquinoline) (Alq_3, Fig. 4.5a). In 1990, Friend and co-workers showed that it is also possible to fabricate OLEDs using spin-coating deposited polymers, such as poly-phenylene-vinylene (PPV). The structure of a typical so-called "bottom emitting" OLEDs is as follows:

- substrate (glass or a transparent polymer),
- transparent anode layer (e.g. ~150 nm indium-tin-oxide, ITO),
- the active organic semiconductor layer stack (~80–120 nm),
- cathode (e.g. an aluminium layer of typically 100 nm with a few nanometers of LiF, Ba or Ca layer at the interface in order to improve the electron injection),

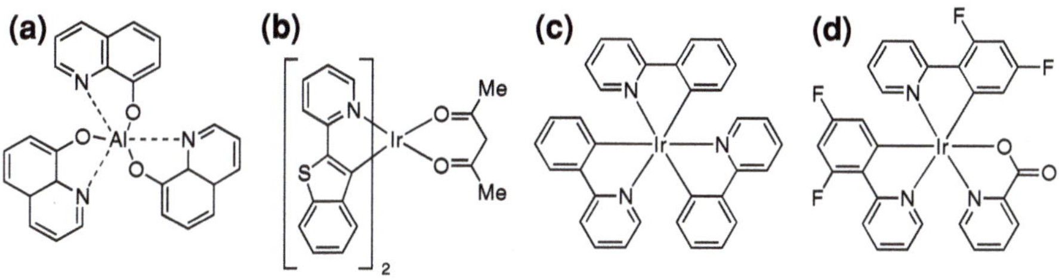

Fig. 4.5 Examples of organic materials used in OLEDs. **a** aluminium-tris-(8-hydroxy-quinolin) (Alq_3, EML and ETL, see Fig. 4.6). **b** iridium(III) bis(2-(2′-benzothienyl) pyridinato-M,C3′) (acetylacetonate) ($btp_2Ir(acac)$, red phosphorescent dopant). **c** fac-tris(2-phenylpyridine)iridium ($Ir(ppy)_3$, green phosphorescent dopant). **d** bis(2-(4,6-difluorophenyl)pyridyl-N,C2′)iridium (acetylacetonate) (FIrpic, blue phosphorescent dopant)

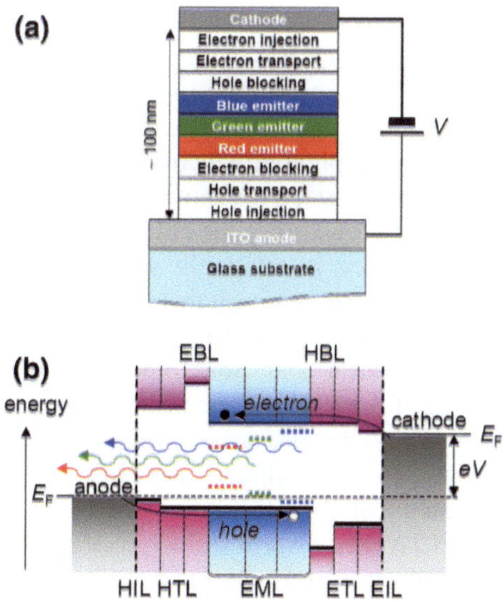

Fig. 4.6 a Schematic structure of a multilayer small-molecule bottom emitting OLED. **b** Schematic energy diagram showing the functioning of the OLED. It contains hole and electron injection layers (*HIL, EIL*), hole and electron transport layers (*HTL, ETL*), electron and hole blocking layers (*EBL, HBL*), and an emissive layer (*EML*) containing *red, green* and *blue* emitting dye molecules (with LUMO and HOMO energies indicated by *dashed lines*). The figure shows the "flat band" situation (neglecting the effects of space charge in the layers on the actual position dependence of the HOMO and LUMO levels). E_F is the Fermi energy, and V is the applied voltage

- a metal, a glass lid or a thin film encapsulating layer structure which protects the cathode from the air and water in the ambient atmosphere.

The active organic semiconductor layer stack can, in principle, consist of a single organic layer containing a mixture of molecules emitting at different wavelengths. However, in the most efficient white OLEDs which are presently developed for general lighting applications, the active organic layer consists of many sublayers, each fulfilling a specific function. A typical example is given in Fig. 4.6a.

Figure 4.6b gives the energy landscape in the OLED at a voltage equal to the built-in voltage, V_{bi}, which is the effective work function difference between the anode and the cathode. The figure shows, firstly, the injection of electrons and holes from the Fermi level energy (E_{F}) in the cathode and anode, respectively, into the lowest unoccupied molecular orbital (LUMO) states and the highest occupied molecular orbital (HOMO) states of the organic semiconductors used (curved arrows). The HOMO and LUMO states are separated by an energy gap (white zone). Efficient electron and hole injection can only take place from electrode layers with a low and a high effective work function, respectively. For simplicity, the figure neglects the effects of the presence of the injected space charge in the OLED, which is in actual devices quite important, causing "band bending". For voltages $0 < V < V_{\mathrm{bi}}$, only a very small current can flow, due to charge carrier diffusion. In a properly designed OLED, with small injection barriers, the OLED "turns on" at $V \approx V_{\mathrm{bi}}$. The Coulomb attraction between the injected electrons and holes leads to the formation of bound electron–hole pairs (excitons), which can recombine radiatively to form a photon, which gives rise to light emission.

The zone within which the emission takes place is confined to an emissive layer (EML) in the center, in order to maximize the efficiency of light-outcoupling out of the microcavity. This efficiency is relatively small for emission close to the electrodes, due to fast irreversible energy transfer to the electrodes. For this purpose, use is made of a combination of electron and hole transport layers (ETL and HTL, respectively) and electron and hole blocking layers (EBL and HBL, respectively) at either side of the EML. The emissive layer (EML) consists usually of at least three sublayers, giving rise to red, green and blue emission, so that overall white emission at a color coordinate very close to the black body curve is obtained with a correlated color temperature which depends on the required application (varying from 2700 K, "warm white", to above 4000 K, "cold white") and with a high CRI. The electron and hole transport layers separate the EML from the electrodes, and should have a large electron and hole mobility, respectively, in order to assure that the voltage drop is confined mainly to the EML. Efficient injection into the transport layers can be obtained by making use of *n*-doped and *p*-doped electron and hole injection layers (EIL and HIL) at the anode and cathode interfaces, respectively [11]. The conductivity of these layers is very high, resulting in a negligible voltage drop.

The efficiency of the emission within each sublayer within the EML depends, in the first place, on the quantum–mechanical nature of the excitons formed. In organic materials based on only light atoms, such as carbon, oxygen, nitrogen and hydrogen, one may strictly distinguish singlet excitons (total spin quantum number $S = 0$) and triplet excitons ($S = 1$). Radiative decay of singlet excitons to the ($S = 0$) ground state, called *fluorescence*, is a quantum-mechanically allowed process and occurs usually within 1–10 ns. In contrast, radiative decay of triplet excitons to the ground state, called *phosphorescence*, is a quantum-mechanically forbidden process. From quantum statistics, it is expected that exciton formation between electrons and holes with uncorrelated spins leads to singlet and triplet excitons in a ratio of 1:3. In OLEDs based on fluorescent emitters, the emission efficiency is therefore severely limited by the small singlet fraction, $\eta_S = 0.25$. In fluorescent emitters, the triplet exciton lifetime is very long, sometimes more than 1 ms. Almost all triplets decay non-radiatively after diffusion to quenching sites. Experimental studies suggest that η_S is indeed close to 0.25 for small-molecule materials based on only light elements, but it has been argued that it is significantly larger for some polymers [12, 13].

The OLED efficiency can be increased drastically by making use of organic materials in which heavy metal atoms such as Ir or Pt are present, in which spin–orbit interaction gives rise to a relatively strong admixture of singlet character into otherwise predominantly triplet states [14]. In such "phosphorescent emitters" or "triplet-emitters", such as the Ir-dye molecules shown in Fig. 4.5b–d, the radiative lifetime is decreased to typically 1 μs, and efficient emission from excitons with predominant singlet *and* triplet character is obtained. The phosphorescent molecules are often present as red, green and blue dopants in a host material with a high energy gap. The corresponding HOMO and LUMO energy levels are indicated as dashed lines in Fig. 4.6b.

For fluorescent OLEDs, the external quantum efficiency (EQE, η_{EQE}, defined as the total number of emitted photons as measured in an integrating sphere spectrometer [15] per electron passing the device) is determined by four factors:

$$\eta_{EQE} = \eta_{ef}\eta_S\eta_{rad}\eta_{lo}. \tag{4.1}$$

Here, η_{ef} is the exciton formation efficiency, i.e. the fraction of the injected carriers which actually form an exciton, η_S is the singlet fraction discussed above, η_{rad} is the singlet exciton radiative decay probability and η_{lo} is the light-outcoupling efficiency. For phosphorescent OLEDs, within which all excitons can emit, $\eta_{EQE} = \eta_{ef}\eta_{rad}\eta_{lo}$. In multilayer OLEDs containing electron and hole blocking layers, η_{ef} is very close to 1. The radiative decay probability, which is limited by the diffusion of excitons to quenching sites, can be maximized by appropriate chemical synthesis and purification procedures. This efficiency, which is often taken to be equal to the photoluminescence quantum efficiency (η_{PL}), is for certain materials close to 1. For phosphorescent OLEDs with perfect blocking layers, η_{EQE} may at small voltages be limited only by the light-outcoupling efficiency, as has been demonstrated in particular for green-emitting OLEDs based on $Ir(ppy)_3$ as the emitting dye molecule [16]. This implies that η_{rad} is then very close to 1. At high voltages and current densities, a roll-off of the efficiency is often observed, which can be attributed to a reduction of η_{rad} due to various exciton quenching processes, such as triplet–triplet exciton annihilation, singlet-singlet exciton annihilation, singlet-polaron and field-induced quenching, depending on the material system involved, and to a reduction of η_{ef} (less efficient charge carrier blocking) [17].

For the case of monochromatic emission at a wavelength λ (with a corresponding photon energy $h\nu$, where h is the Planck constant and ν the photon frequency), the luminous efficacy at a voltage V is given by

$$\eta_P(V) = \eta_{EQE}(V)K(\lambda)\frac{h\nu}{eV} \quad \left(\text{in units lm W}^{-1}\right). \tag{4.2}$$

Here $K(\lambda) = 683V_{\text{eye-sense}}(\lambda)$, with $V_{\text{eye-sense}}(\lambda)$ is a dimensionless eye-sensitivity function, which is equal to 1 at the maximum of the eye sensitivity curve for $\lambda = 555.17$ nm (yellow–green). Generalization of Eq. (4.2) for the case of emission with a broad spectrum shows that the maximum possible luminous efficacy of a high-quality white light source is approximately 350–400 lm W^{-1}, depending on the color point and the CRI. In that case, the applied voltage should be very close to the (highest emitted) photon energy.

Using a full phosphorescent layer structure ("PPP", i.e. phosphorescent (P) red, green and blue layers), the internal quantum efficiency can be close to 100 %. For a planar OLED with randomly oriented emitting dipoles, the outcoupling efficiency is typically 20 %, so that in this case the EQE is also 20 %. Highly stable blue phosphorescent emitters are not yet widely available. Therefore, an alternative route towards efficient white OLEDs in which blue emission is obtained from fluorescent (F) molecules ("PPF" layer structure), has attracted much interest. For such "hybrid" OLEDs, optimization of the EQE up to 15 % and improved lumen maintenance (lifetime) have been demonstrated by Loebl et al. [1]. This result was achieved without methods for improved light-outcoupling. In order to prevent transfer of the singlet excitons in the blue emitting layer to the adjacent green or red emitting layers, a thin ($\sim$5 nm) interlayer which serves as an exciton blocking layer is then often used. The (singlet) exciton energy in the interlayer should be relatively high, whereas it should still function as an ambipolar (electron and hole) transport layer.

The efficiency of hybrid OLEDs can be enhanced by making use of a layer concept within which the triplet excitons that are generated in the blue fluorescent layer are harvested by letting them diffuse to the adjacent red and/or yellow emitting layers [18, 19]. Segal et al. have proposed another concept for harvesting the triplet excitons in a blue emitting fluorescent layer, called "extra-fluorescence" [20]. Use is made of an adjacent layer containing blue-phosphorescent dye molecules, which convert triplets in the fluorescent layer to singlets (intersystem crossing) due to the temporary formation of a charge-transfer exciton (exciplex) at the interface between both layers. Although an efficiency enhancement was found using this method, improvements beyond the state-of-the-art have not yet been demonstrated. From this work, it follows that the optimal design of OLED layer stacks does not only require optimizing the charge carrier balance, in order to generate excitons in appropriate ratios in the R, G and B layers, but also carefully optimizing the exciton transfer processes.

For creating white OLEDs, several layer concepts can be used (Fig. 4.7).

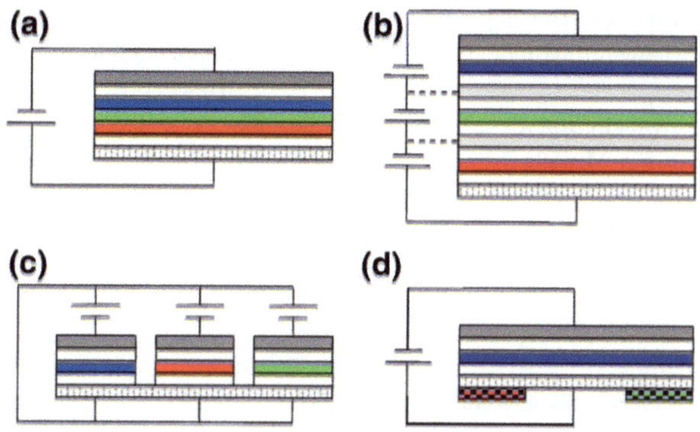

Fig. 4.7 a–d Methods for obtaining white emission, as explained in the text. The *colored* and *white layers* are emissive and transport layers, respectively. The *dark grey* and *dotted white layers* are the cathode and anode, respectively. In figure **d** external *red* and *green* phosphorescent layers are present

- The focus in this paper has been on OLEDs within which R, G and B emissive layers form together a central emissive layer (Fig. 4.7a). The emitters need not necessarily be part of separate layers. Recently, efficient white emission (EQE = 18 %) has been reported from an OLED within which co-doped phosphorescent blue and red emitters are present in a high-energy gap matrix material [21].
- Alternatively, a stack structure may be used within which individual R, G and B emitting OLEDs are combined, separated by strongly doped charge conversion layers within which an electron current can be converted in a hole current. When a thin transparant metal layer is used instead, in order to enhance the lateral conductivity, connection of that layer to a voltage supply can even make it possible to obtain color tunability (Fig. 4.7b).
- Color tunability is, in principle, also possible by making use of a pixelated OLED, such as in a display (Fig. 7c).
- White emission can also be obtained from a blue OLED, on top of which external green and red wavelength-shifting phosphorescent layers are added (Fig. 7d).

In Table 4.1, the development of the luminous efficacy, the operational lifetime and the type of materials used for the red, green and blue emitting layers as obtained in various laboratories and within the EU-funded project OLLA [45] is given. The operational lifetime is in general defined as the time at which the brightness of an OLED, initially at 1000 cd m^{-2}, has dropped by a factor of two (see further below). The most important factor limiting the luminous efficacy of fully planar white phosphorescent multilayer OLEDs is the light-outcoupling efficiency, which is typically 20 % as mentioned already above. An enhancement of the light-outcoupling by a factor up to 1.5 can be achieved by roughening of the glass substrate, e.g. by adding a rough foil as shown schematically in Fig. 4.8. A much larger enhancement can be obtained by making use of so-called macro-extractors (glass-spheres), in order to obtain emission from modes which would otherwise stay in the glass due to internal wave guiding. For normal glass substrates (with standard refractive index n), the enhancement factor can be 2. For special (high-index substrates), the enhancement can be up to a factor 3 [22]. Of course, the use of such

Table 4.1 Development of white OLEDs

Company/Institute	Year	Luminous efficacy[a] (lm/W)	Lifetime[a] (h)	Emitter type R,G,B
Novaled/Philips	2006	32	20.000	P, P, F
Konica-Minolta	2006	64	10.000	P, P, P
The OLLA-project	2007	25	5.000	P, P, F
Idemitsu Kosan	2007	17[b]	30.000[b]	F, F, F
Osram	2008	46	5.000	P, P, F
Novaled	2008	35	100.000	P, P, F
The OLLA-project	2008	51	>10.000	P, P, F
Philips	2008	48 (80)*	>10.000	P, P, F
Universal display corporation	2008	(102)*	8.000	P, P, P
BASF/Osram	2008	60	–	P, P, P
Panasonic	2009	40	20.000	P, P, F

[a] Measured at 1000 cd m^{-2} , unless stated otherwise

[b] Measured at 10 mA/cm^2

* Measured using a glass hemisphere or a macro-extractor in order to obtain enhanced light outcoupling by extracting light which would otherwise be waveguided in glass modes

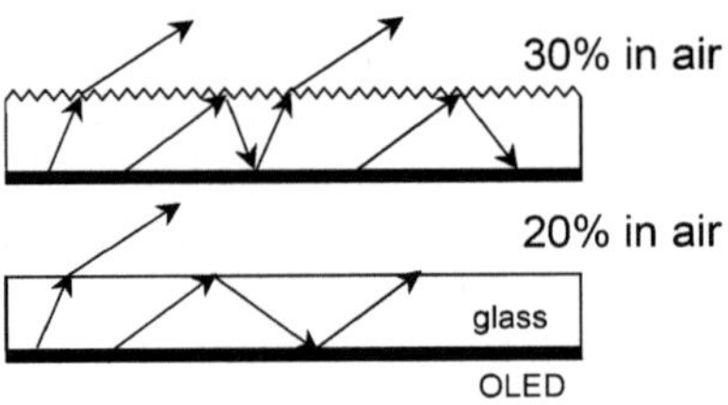

Fig. 4.8 Schematic view of the emission enhancement from an OLED of which the glass substrate has been roughened (*top figure*), as compared to the emission from an OLED with a fully planar glass substrate (*bottom figure*). The use of a roughened surface can increase the light-outcoupling efficiency from approximately 20 to 30 %

special elements is not wanted for general lighting applications, as the main features of an OLED are lost.

We note that deviations of the color from the black-body curve can also give rise to enhanced luminous efficacy. However, for applications as a white light source only minor deviations are acceptable. Furthermore, various methods for improved outcoupling have been proposed using non-flat internal interfaces, or even using emission from lithographically created micrometer-scale emissive regions which are separated by non-emissive low-index regions [23]. Other efficiency limiting factors include the use of fluorescent instead of phosphorescent blue emitting layers (in view of a lack of widely available highly stable blue phosphorescent emitters), and the efficiency roll-off (at higher brightnesses) due to the factors already discussed above.

The operational lifetime is determined by the ability to maintain the lumen output over time: the light emitting materials degrade slowly over time, leading to a gradual reduction in light emission. The degradation rate strongly depends on the type of material, and recent developments have strongly improved this operational lifetime, as illustrated in Table 4.1. This lifetime, or lumen maintenance, is typically expressed as the time until the brightness has dropped to a certain percentage of the original brightness at $t = 0$. Often, a percentage of 50 % is used (the corresponding lifetime is indicated as LT50), although from an application point of view, a percentage of 70 or even 80 % would be more appropriate (LT70 or LT80). For fair comparison, the data presented in the table are all measured at a brightness of 1000 cd/m^2, unless indicated otherwise: a lower brightness will usually lead to (much) longer lifetimes, whereas higher brightness levels will shorten the lifespan of the device.

In order to achieve further improvements of the luminous efficacy and lifetime, the availability of quantitative numerical electronic and optical device models is expected to be helpful. Modeling of the light-emission efficiency is based on microcavity models which treat the emission in a mixed classical and quantum–mechanical way [24, 25]. Accurate knowledge of the complex refractive indices of all layers involved and of the emissive dipole orientation in the emitting layers is required. Advanced models include the effects of lateral variations of the optical cavity, which are introduced to enhance the light-outcoupling efficiency [26]. Modeling of the electrical transport and recombination requires the inclusion of the effects of the energetically disordered nature of the HOMO and LUMO states on the mobility. Quantitative device models which take this effect into account frequently assume a Gaussian shape of the density of states ("Gaussian Disorder Model", [27–29]), in some cases with a superimposed exponential low-energy tail and in the case of the emissive layers including the HOMO or LUMO states due to the dye molecules. In such cases, the mobility increases not only with increasing electric field, but also with increasing local charge carrier density. As a refinement, spatially correlated energetic disorder can be taken into account [30, 31]. These models have been most intensively applied to hole and electron transport in polymers [32], but have recently been found to be relevant as well to small-molecule materials [33]. Three-dimensional calculations show that as a result of the energetic disorder the current density in OLEDs can be quite filamentary [34, 35]. Understanding the filamentarity of the current density, potentially leading to recombination which is quite localized at preferential "hot-spots", might be of importance in order to better understand and control lifetime limiting processes. Various software tools based on one-dimensional simulations of the light-outcoupling, electrical transport and recombination are presently available commercially (SETFOS, from the company Fluxim, and simOLED from the company sim4tec).

4.3 Fabrication Technologies

As described in the previous section, OLEDs typically have a complex layer stack, comprising several different and patterned layers, sometimes as many as 20 or even more. Because of their sensitivity to moisture, OLEDs also need to be

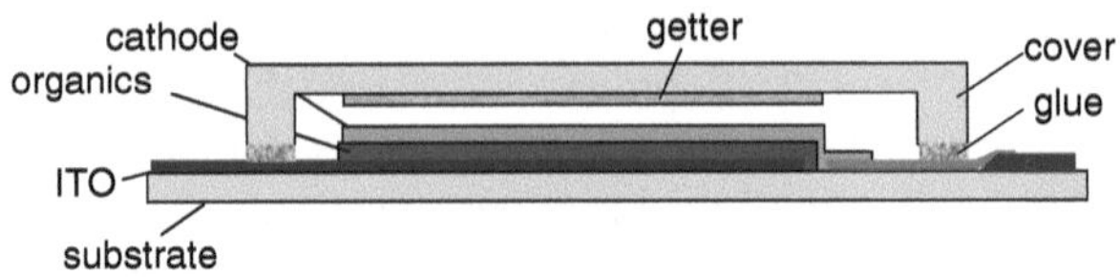

Fig. 4.9 Schematic cross-section of an encapsulated OLED device. Light is emitted through the transparent ITO anode and the substrate

Fig. 4.10 Standard process flow, illustrating the various steps: **a** structured ITO. **b** Metal contacts. **c** Patterned isolation. **d** Light emitting organics. **e** Local removal of organics (polymer only). **f** Cathode. **g** Encapsulation

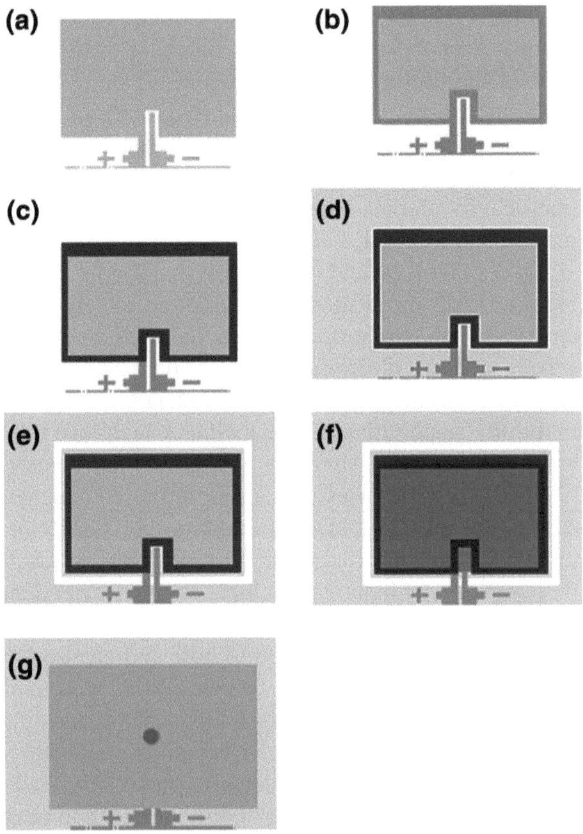

hermetically sealed (encapsulated, see Sect. 4.4). This poses a serious challenge for cost effective manufacturing of OLED products. The following paragraphs outline the current manufacturing technology and an outlook towards roll-to-roll production.

The majority of the research and (pre)pilot lines currently installed is based on technology similar to that used in the flat panel display fabrication, with the obvious exception of the thin film transistor (TFT) backplane. The process flow for a device such as shown in Fig. 4.9 is described step by step in Fig. 4.10.

The process starts with a substrate of glass, typically soda-lime, ranging in size from 6” to Gen 2.5 (40 × 50 cm^2). First layer (Fig. 4.10a) is the transparent anode, often ITO (indium tin-oxide), which is sputtered and annealed, and structured using a photolithographic mask followed by etching.

Next is a metal layer (Fig. 4.10b) of about 300 nm, for example Cr/Al to make contact pads and bus bars for current distribution around the device. For larger device areas (smallest dimension > few cm), shunt lines are required as well to lower the effective sheet resistance of the anode: without such shunt lines, the light output of the OLED will be lower in the centre than at the edges. This metal layer is either sputtered or evaporated using a shadow mask for patterning, or again using photolithography. To prevent a direct contact between anode and cathode, a thin layer of an electrical isolator is deposited, often a patterned photoresist (Fig. 4.10c).

In case of a small-molecule device, the materials of the various layers are evaporated, often through a shadow-mask. Alternatively, deposition from the gas phase using a carrier gas may be used (Organic Vapor Phase Deposition, OVPD [36]). However, this is not the standard process. Polymer light emitting material can be deposited by spin coating, followed by a laser ablation step around the edges to prevent moisture ingress (Fig. 4.10d,e).

The last layer is the cathode, typically 100 nm of evaporated Al (Fig. 4.10f). Again, a shadow mask is used to restrict deposition to the active area. Finally, either a metal or glass lid comprising a desiccant material (getter) is glued to the OLED to encapsulate the device (Fig. 4.10g).

Realizing uniformity of the emission of large-area OLEDs requires that the electrodes function, effectively, as equipotential planes. The effective device resistance (V/I) should therefore be much larger than the sheet resistance of the anode and cathode layers. For typical OLEDs with a $\sim$50–100 nm ITO anode layer, with a sheet resistance of the order 10–50 Ω, this condition is in general only met for devices smaller than 1 cm^2. Uniform emission from larger areas can be achieved by making use of a conducting grid structure [37].

In order to meet the cost targets for general lighting applications, the process flow sketched above will need significant changes. The number of (expensive) processing steps such as the photolithography must be reduced, for example using additive processes such as printing. High throughput systems will lower the impact of machine costs, while alternative materials such as ITO replacements are expected to further bring down the costs. A promising approach currently being investigated is roll-to-roll production of OLEDs on low-cost, flexible substrates such as PET or PEN. The use of flexible substrates opens up a host of new possibilities in terms of processing options, but also in applications areas. However, this requires a new type of encapsulation which is flexible, low cost but still hermetic. Flexible substrates such as PET or PEN are quite permeable to water vapor, and require a transparent barrier layer which is as hermetic as the flexible encapsulation. This 'thin-film encapsulation' will be discussed in more detail in Sect. 4.4.

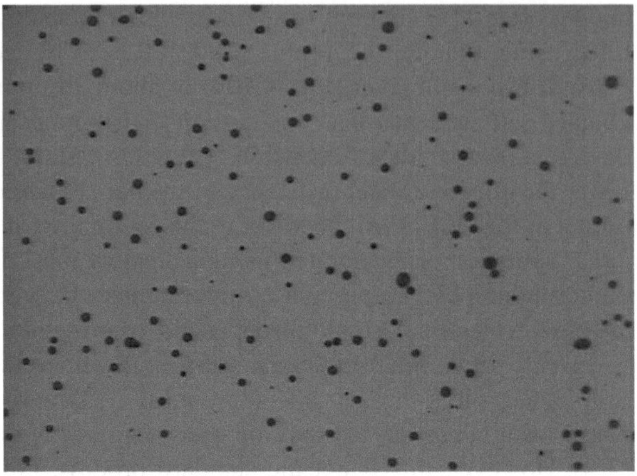

Fig. 4.11 Example of the formation of *black spots* in electroluminescence of an OLED as a result of storing the device at ambient conditions for a few days. The area of the picture is approximately 3 mm^2

4.4 Encapsulation

Exposure to the ambient atmosphere of OLEDs which are not properly packaged would result in a reduction of the practical lifetime of the device, during shelf life or operation. The most pronounced failure as a result of this interaction is the formation of so-called black spots which can be observed in electroluminescence (Fig. 4.11). A qualitative description of the mechanism of this formation is reported in the literature [38, 39]: water from the ambient atmosphere penetrates through pinholes in the cathode layer, leading to oxidation of metal at the cathode interface, preventing electron injection during operation of the device. This is observed as a region without emission, i.e. a black spot. Because the oxidation process is determined by water diffusion from the pinhole, the black spot will be spherical in shape and its area grows linearly with time. This process is a shelf effect, i.e. no current or voltage is necessary to drive it.

Clearly, for any application, black spots should be avoided or at least suppressed, as they affect the visual appearance and continuously shrink the light emitting area. To prevent black spots, OLEDs require extremely good encapsulation to prevent oxidation of the cathode. Conventionally, this is done with a metal or glass lid on top of the device, comprising a desiccant getter material such as CaO to remove any residual moisture. This lid is glued onto the glass OLED substrate using an epoxy resin (see Fig. 4.9). Although lower-cost solutions will be needed in future, this approach works well in preventing black spot formation for many years.

When fabricating OLEDs on flexible substrates this standard approach is not suitable. Instead of rigid lids, thin film barriers consisting of inorganic and also inorganic/organic multi-layers are often used [40, 41]. This so-called thin film encapsulation is applied directly on top of the cathode (Fig. 4.12), with the purpose to cover the pinholes in the aluminium layer. To be an effective barrier, the water vapour transmission rate (WVTR) needs to be extremely low, in the order of 10^{-6} g/m^2/day or better. Such low WVTR values can be obtained already by a single layer of SiN,

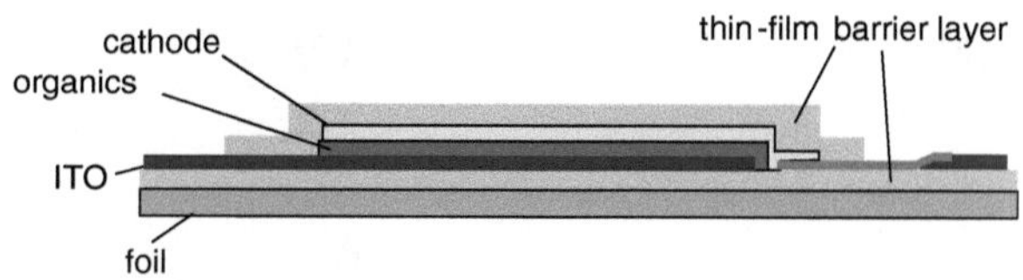

Fig. 4.12 Schematic diagram of a flexible OLED device that is protected by a thin-film barrier layer on both sides of the OLED device

deposited using a plasma enhanced chemical vapour deposition (PECVD) process under conditions that are compatible with the OLED processing conditions. However, one such layer only covers 90–99 % of the pinholes in the cathode, still leaving several uncovered pinholes per cm^2. Simply increasing the SiN thickness only helps to a limited extent and is not a practical solution. Therefore, stacking of multiple, often alternating layers is a widely used approach. The intermediate layers, either inorganic such as SiO_2 or organic such as polyacrylate help in decoupling the pinholes in the successive layers, leading to a much improved coverage and thus better barrier performance. State-of-the-art encapsulation results in significant yield of $2 \times 2\ cm^2$ OLEDs without black spots for 500 h at 60 °C and 90 % relative humidity, corresponding to about 10 years at ambient conditions [42].

In order to measure the low water vapor transmission rate of the applied barrier layer that is relevant for a long lifetime of an OLED device, the so-called calcium test has been developed: a pattern of calcium of well-defined thickness (typically 50 nm) is applied on a transparent substrate, and the calcium is covered with a transparent encapsulant. During the test period, the optical transmission of the calcium layer is measured with a lamp as a light source and a camera as detector: calcium as a metal has a low transmission, but after reaction with water or oxygen the layer becomes transparent. Measurement of the optical transmission as a function of time thus provides information of the permeability of the encapsulant (and substrate) for water and oxygen. The practical detection limit of the calcium test is in the order of $10^{-6}\ g/m^2/day$. However, as described in a recent white paper on thin film barrier characterisation [43], an OLED device is two orders of magnitude more sensitive to water than a calcium test. This huge difference is related to the amount of material to be oxidized before a detectable spot has formed: 50 nm for a 'white spot' in a calcium test versus sub-nm for a black spot in an OLED. Clearly, results from a calcium test cannot be compared directly to results based on the formation of black spots in an OLED. The only relevant test in the qualification of an encapsulant is a test on OLED devices with sufficient area (at least 10 cm^2), where black spots should not become visible within the application-specified period during a shelf test.

4.5 Summary and Outlook

OLED technology is presently developing as a promising option for large area lighting applications, with basic properties such as efficiency, color stability and lifetime which approach those of conventional lighting technology, and with

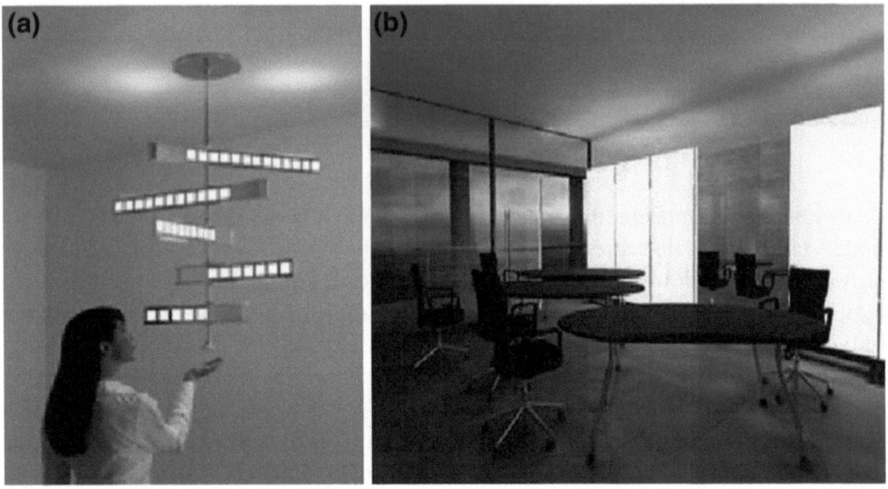

Fig. 4.13 Special-lighting (**a**) and general-lighting (**b**) OLED applications. From: www.the-new-art-of-light.com/

various interesting additional complementing features. The first applications of OLED lighting are expected in the specialized markets, emphasizing the novel design opportunities. Later developments will make it possible to realize general lighting (Fig. 4.13), such as office lighting. Transparent OLEDs are expected in 3–5 years, whereas flexible OLEDs (light foils) such as shown in Fig. 4.3 are expected in 5–8 years. Recent academic developments suggest that even more fascinating OLED structures, which conformally cover three-dimensional surfaces, could become feasible [44].

The progress in OLED technology is in part a result of intensive collaborations between research in academia, institutes, OLED-producing companies and supplying companies, e.g. of chemical components and fabrication tools. In Europe, strong contributions on efficient white OLEDs for general lighting have been made by EU-funded projects such as OLLA (2005–2008) [45] and OLED100 (2009–2012) [46]. The goal of the latter project is to develop the technologies for obtaining 100 lm/W luminous efficacy, 100,000 h lifetime, and $100 \times 100\ \text{cm}^2$ size. More specific contributions are made by the projects Fast2Light [43], which aims at developing novel cost effective high-throughput roll-to-roll large area deposition processes for fabricating light-emitting polymer-OLED foils for intelligent lighting applications, and by AEVIOM [47], which aims at developing a fast and user-friendly integrated electrical/optical OLED model with predictive quality, as a means to enable breakthroughs in white OLED efficiency, lifetime, and manufacturing cost.

Acknowledgments The authors wish to thank P. van de Weijer and S. Grabowski for useful comments. This research has received funding from the European Community's Program No. FP7–213708 (AEVIOM, contribution R.C.).

References

1. Peter Loebl, Volker van Elsbergen, Herbert Boerner, Claudia Goldmann, Stefan Grabowski, Dietrich Bertram (2009) White OLEDs for lighting applications, Proceedings of SPIE, Optics and Photonics 2009: Photonic devices and applications, San Diego August 2–6, 7415:74151A–1
2. D'Andrade BW, Forrest SR (2004) White organic light-emitting devices for solid state lighting. Adv Mater 16:1585
3. Shinar ZJ (2004) Organic light-emitting diodes—a survey. Springer, New York
4. Kalinovski J (2005) Organic light-emitting diodes: principles characteristics and processes. Marcel Dekker, New York
5. Brütting W (ed) (2005) Physics of organic semiconductors. Wiley, Weinheim
6. Mullen K, Scherf U (2006) Organic light emitting devices: synthesis properties and applications. Wiley, Weinheim
7. Li ZR, Meng H (eds) (2007) Organic light-emitting materials and devices. Taylor and Francis, Boca Raton
8. So F, Kido J, Burrows P (2008) Organic light-emitting devices for solid-state lighting. MRS Bulletin 33:663–669
9. Pope M, Svenberg CE (1982) Electronic processes in organic molecular crystals. Oxford University Press, New York
10. Tang CW, VanSlyke SA (1987) Organic electroluminescent diodes. Appl Phys Lett 51:913
11. Pfeiffer M, Leo K, Zhou X et al (2003) Doped organic semiconductors: physics and application in light emitting diodes. Org Electron 4:89–103
12. Cao Y, Parker ID, Yu G, Zhang C, Heeger AJ (1999) Improved quantum efficiency for electroluminescence in semiconducting polymers. Nature 397:414–417
13. Wilson JS et al (2001) Spin-dependent exciton formation in p-conjugated compounds. Nature 413:828–831
14. Baldo MA, O'Brien DF, You Y et al (1998) Highly efficient phosphorescent emission from organic electroluminescent devices. Nature 395:151–154
15. Forrest SR, Bradley DDC, Thomson ME (2003) Measuring the efficiency of organic light-emitting diodes. Adv Mater 15:1043
16. He G, Pfeiffer M, Leo K, Hofmann M, Birnstock J, Pudzich R, Salbeck J (2004) High-efficiency and low-voltage p-i-n electrophosphorescent organic light-emitting diodes with double-emission. Appl Phys Lett 85:3911
17. Giebink NC, Forrest SR (2008) Quantum efficiency roll-off at high brightness in fluorescent and phosphorescent organic light emitting diodes. Phys Rev B 77:235215
18. Sun Y, Giebink NC, Kanno H, Thompson ME, Forrest SR (2006) Management of singlet and triplet excitons for efficient white organic light-emitting devices. Nature 440:908–912
19. Schwartz G, Fehse K, Pfeiffer M, Walzer K, Leo K (2008) Reduced efficiency roll-off in high-efficiency hybrid white organic light-emitting diodes. Appl Phys Lett 92:053311
20. Segal M, Singh M, Rivoure K, Difley S, Van Voorhis T, Baldo MA (2007) Extrafluorescent electroluminescence in organic light-emitting devices. Nature Mater 6:374
21. Jeon SO, Yook KS, Joo CW, Lee JY (2009) Highly efficient single-layer phosphorescent white organic light-emitting diodes using a spirofluorene-based host material. Opt Lett 34:407
22. Reineke S et al (2009) White organic light-emitting diodes with fluorescent tube efficiency. Nature 459:234–239
23. Sun Y, Forrest SR (2008) Enhanced light-outcoupling of light-emitting devices using embedded low-index grids. Nat Photonics 2:483
24. Lu MH, Sturm JC (2002) Optimization of external coupling and light emission in organic light-emitting devices: modeling and experiment. J Appl Phys 91:595
25. Bulovic V, Khalfin VB, Gu G, Burrows PE, Gharbuzow DZ, Forrest SR (1998) Weak microcavity effects in organic light-emitting device. Phys Rev B 58:3730

26. Greiner H (2004) Exploring particlelike nanostructures for light outcoupling from organic LEDs by first principles calculations. Proc SPIE 5450:376–387
27. Bässler H (1993) Charge transport in disordered organic photoconductors - A Monte Carlo simulation study. Phys Stat Sol B 175:15
28. Pasveer WF et al (2005) Unified description of charge-carrier mobilities in disordered semiconducting polymers A. Phys Rev Lett 94:206601
29. Coehoorn R, Pasveer WF, Bobbert PA, Michels MAJ (2005) Charge-carrier concentration dependence of the hopping mobility in organic materials with Gaussian disorder. Phys Rev Lett 94:206601
30. Gartstein YN, Conwell EM (1995) High-field hopping mobility in molecular systems with spatially correlated energetic disorder. Chem Phys Lett 245:351
31. Bouhassoune M, van Mensfoort SLM, Bobbert PA, Coehoorn R (2009) Carrier-density and field-dependent charge-carrier mobility in organic semiconductors with correlated Gaussian disorder. Org Electr 10:437
32. van Mensfoort SLM, Vulto SIE, Janssen RAJ, Coehoorn R (2008) Hole transport in polyfluorene-based sandwich-type devices: Quantitative analysis of the role of energetic disorder. Phys Rev B 78:085208
33. van Mensfoort SLM, Shabro V, de Vries RJ, Janssen RAJ, Coehoorn R (2010) Hole transport in the organic small-molecule material a-NPD: evidence for the presence of correlated disorder. J Appl Phys 107:113710
34. Tutis EE, Batistic I, Berner D (2004) Injection and strong current channeling in organic disordered media. Phys Rev 70:161202(R)
35. van der Holst JJM et al (2009) Modeling and analysis of the three-dimensional current density in sandwich-type single-carrier devices of disordered organic semiconductors. Phys Rev B 79:085203
36. Shtein M, Gossenberger HF, Benzinger JB, Forrest SR (2001) Material transport regimes and mechanisms for growth of molecular organic thin films using low-pressure organic vapor phase deposition. J Appl Phys 98:1470
37. Neyts K, Real A, Marescaux M, Mladenovski S, Beeckman J (2008) Conductor grid optimization for luminance loss reduction in organic light emitting diodes. J Appl Phy 103:093113–093115
38. Berntsen AJM et al (1998) Stability of polymer light-emitting diodes. Philips J Res 51:511–525
39. Kim J-S et al (2002) Nature of non-emissive black spots in polymer light-emitting diodes by in-situ micro-Raman spectroscopy. Adv Mater 14:206–209
40. Burrows PE et al (2001) Ultra-barrier flexible substrates for flat panel displays. Display 22:65–69
41. Young ND et al (2003) Low temperature poly-Si on flexible polymer substrates for active matrix displays and other applications. Proc MRS 769:17–29
42. T. v. Mol (2008) Flexible barrier films. N. Holst Centre, (ed.) Berlin: 4th International plastic electronics conference and showcase 2008
43. www.fast2light.eu
44. Köhnen A et al (2009) The simple way to solution-processed multilayer OLEDs—layered block-copolymer networks by living cationic polymerization. Adv Mat 21:879
45. www.hitech-projects.com/euprojects/olla/
46. www.oled100.eu
47. www.aeviom.eu

Chapter 5
Large Area Electronics with Organic Transistors

Makoto Takamiya, Tsuyoshi Sekitani, Koichi Ishida, Takao Someya and Takayasu Sakurai

Abstract Organic electronics is attracting a lot of attention for large-area pervasive electronics applications, because organic transistors can be fabricated using printing technologies on arbitrary substrates and this enables both high-throughput and low-cost production. In this chapter, some examples of large area electronics based on organic transistors including an EMI measurement sheet, a wireless power transmission sheet, and a communication sheet are presented. Challenges for future large area electronics are also described.

Keywords Large area electronics · Organic transistor · EMI · Power · Communication

5.1 Introduction

Organic electronics is attracting a lot of attention for large-area pervasive electronics applications, because organic transistors can be fabricated with printing technologies on arbitrary substrates, enabling both high-throughput and low-cost

M. Takamiya (✉)
VLSI Design and Education Center, University of Tokyo, 4-6-1 Komaba, Meguro-ku, Tokyo 153-8505, JAPAN
e-mail: mtaka@iis.u-tokyo.ac.jp

T. Sekitani · T. Someya
Department of Electrical and Electronic Engineering, University of Tokyo, 7-3-1 Hongo, Bunkyo-ku, Tokyo 113-8656, JAPAN

K. Ishida · T. Sakurai
Institute of Industrial Science, University of Tokyo, 4-6-1 Komaba, Meguro-ku, Tokyo 153-8505, JAPAN

E. Cantatore (ed.), *Applications of Organic and Printed Electronics*, Integrated Circuits and Systems, DOI: 10.1007/978-1-4614-3160-2_5,

Table 5.1 Comparison between organic FETs (OFETs) developed in our group and the state-of-the-art 45 nm silicon MOSFETs

	OFETs	Si MOSFETs
Minimum gate length	20 μm	45 nm
Mechanical flexibility	Flexible	Very limited
Normalized on current	3 nA/m@3 V	1 mA/m@1 V
Gate delay	0.1 s@3 V	10 ps@1 V
Cost/area	Low	High
Cost/transistor	High	Low
Lifetime	Months	Years

production. In this chapter, some examples of large area electronics using organic transistors are shown.

Section 5.2 describes the advantage of organic electronics over silicon VLSI's and shows the reason why large area applications are suitable for organic transistors. Section 5.3 presents applications of large area electronics including an EMI measurement sheet, a wireless power transmission sheet, and a communication sheet. Section 5.4 describes challenges for future large area electronics and some concluding remarks are given in Sect. 5.5.

5.2 Advantages of Printed Organic Electronics over Silicon VLSI's and Large Area Electronics with Organic Transistors

In order to find suitable applications for organic electronics, the advantages of this technology over conventional electronics should be clarified. Table 5.1 shows a comparison between our organic FETs (OFETs)[1] and state-of-the-art 45 nm silicon MOSFETs. OFETs are flexible, because OFETs are fabricated on a film. Low voltage operation (3 V) is possible already using organic CMOS technology. The gate delay of the OFETs developed in our group, however, is often 10^{10} times larger than that of the silicon MOSFETs, because the minimum gate length of OFETs is 440 times larger than that of the silicon MOSFETs and the on-current of OFETs is 1/300,000 of that of the silicon MOSFETs (mostly due to the much lower mobility of organic semiconductors).[2] The device lifetime of OFETs is several months,

[1] In other sections of this book OFETs are also called OTFTs.

[2] It should be noticed that the performance of organic transistors develops quickly: complementary inverters with a stage delay of 1μs, employing transistor with L=4μm together with p and n materials having a mobility close to $1cm^2/Vs$ have been demonstrated recently [22]. This kind of result, which does not exploit printing techniques but lithography and evaporation, can be considered close to the maximum performance that organic semiconductors can intrinsically offer.

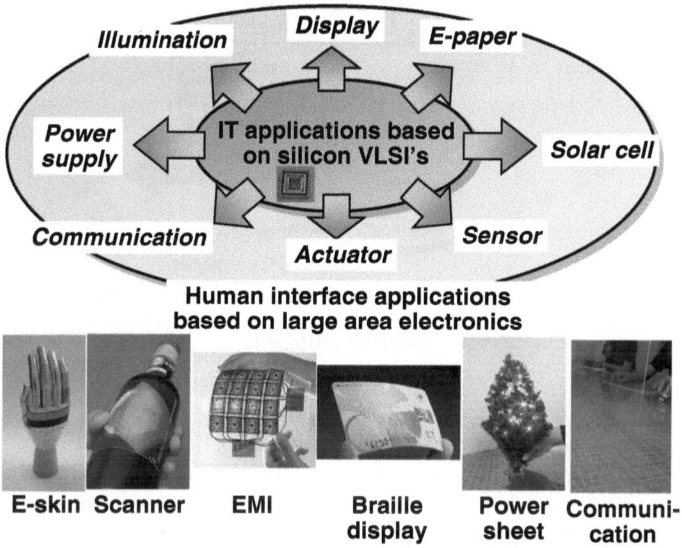

Fig. 5.1 Relation between conventional IT applications using the silicon VLSI and human interface applications using large area electronics. The photos show examples of large area electronics applications. An E-skin [1–3] is a sheet containing an array of pressures sensors and OFETs. A sheet scanner [4–7] is a sheet integrating an array of photodetectors and OFETs. An EMI measurement sheet [8, 9] is a sheet supporting an array of silicon VLSI circuits and OFETs. A Braille display [10–13] is a sheet integrating an array of plastic actuators and OFETs. A wireless power transmission sheet [14, 15] and a communication sheet [16–19] combine on foil an array of coils, plastic MEMS switches, and OFETs. The E-skin, the sheet scanner, and the EMI measurement sheet are examples of sensor applications. The Braille display is an actuator. The wireless power transmission sheet and the communication sheet are possible human-size IT applications

while that of the silicon MOSFETs is more than 10 years, this because OFETs are chemically degraded by the oxygen and moisture in the atmosphere. The cost per area of OFETs is lower than that of the silicon MOSFETs, while the cost per transistor of OFETs is higher than that of the silicon MOSFETs. Therefore, large area electronics, which takes advantage of the flexibility and the low cost per area, is a good application of the OFETs. Intense research and development work is thus focusing on large area devices based on organic materials such as an organic electroluminescence (EL) displays, e-paper, organic EL lighting, organic solar cells, and human-size sensor and actuator devices.

Figure 5.1 shows the relation between conventional IT applications based on silicon VLSI and human interface applications using large area electronics. The IT applications such as computers and communications are supported by the conventional silicon VLSI's, because silicon MOSFETs are fast and the cost per transistor (= cost per function) is low. In contrast, human interface applications such as human interfaces and ambient intelligence can well be supported by large area electronics with OFETs, because OFETs are flexible and the cost per area is low. Thus silicon VLSI and OFETs are not competing technologies but more

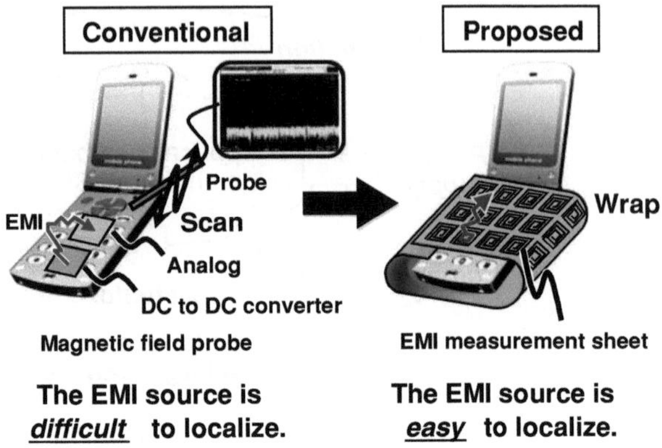

Fig. 5.2 Intrasystem EMC issue in a mobile phone. © 2010 IEEE. Reprinted, with permission, from [9]

complement each other. By combining silicon VLSI's and OFETs, large area electronics has the potential to expand both the application field and the market size of electronic industry. Therefore, the development of the large area electronics which sophisticatedly combines silicon VLSI's and OFETs is strategically very important.

5.3 Applications of Large Area Electronics

5.3.1 EMI Measurement Sheet

Figure 5.2 shows an example of an intrasystem EMC issue in a mobile phone. EMI largely depends on the circuit board layout. Localizing either an EMI source or a critical wiring is difficult by simulation. EMI measurement is, therefore, important for the development of electronic systems. However, there is no method of measuring EMI on the surface of 3D structures. In a conventional method, a pencil-like magnetic field probe with X–Y scanning equipment and spectrum analyzers are used for the EMI measurement [20]. In the method, the surface of the electronic device should be scanned repeatedly with the probe. However, the scanning equipment can only move in a flat plane. In another conventional method, a measurement system with an integrated array of magnetic field loop antennas is used [21]. Although the method captures the distribution of a magnetic field, it is not applicable to 3D structures since the antenna array is implemented on a flat and rigid printed circuit board.

To solve this problem, an EMI measurement sheet [8, 9] was proposed, which enables the measurement of EMI distribution on the surface of 3D structures by wrapping the devices with a sheet like the "furoshiki".[3] Once EMI noise is

[3] A type of traditional Japanese wrapping cloth that is frequently used to transport clothes, gifts, or other goods.

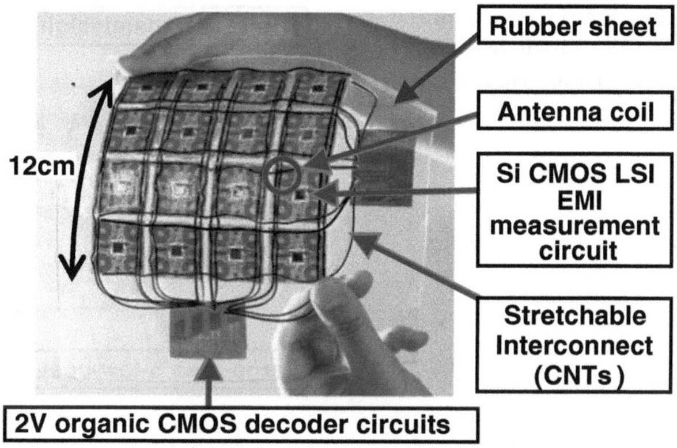

Fig. 5.3 Stretchable EMI measurement sheet. © 2010 IEEE. Reprinted, with permission, from [9]

roughly localized with the measurement sheet, one can easily scan EMI noise using the probe method for precise localization or better quantification of the EM field. The sheet can measure not only a magnetic field but also an electric field suitably changing its antenna connection.

Figure 5.3 shows a prototype of the stretchable EMI measurement sheet. Each printed circuit board (PCB) includes 2 × 2 antenna coils and a silicon CMOS LSI chip. The sheet consists of 4 × 4 PCBs, and therefore, 8 × 8 antennas are located in 12 × 12 cm^2 area. The antennas and LSIs are controlled using 2 V organic CMOS decoder and selector. Each module is electrically connected with a stretchable interconnect made of carbon nanotubes. The overall system is sealed with a rubber sheet made of silicone elastomer. The sheet is, therefore, flexible and stretchable. The sheet detects the total power of an electric field in the band up to 700 MHz and that of a magnetic field up to 1 GHz. The minimum detectable power of the electric and magnetic fields are −60 and −70 dBm, respectively.

5.3.2 Ubiquitous Power Supply and Communication

In future ubiquitous electronics, also known as ambient intelligence, 1,000–10,000 electronic devices will be distributed around the user's environment and will contribute to security, promote healthcare and welfare, and provide entertainment and convenience. In the ubiquitous electronics environment, distributing power supply to the electronic devices and ensuring communication between them are serious problems. This is because battery replacement is not acceptable in this scenario and the power consumption for wireless communication between so many devices will be prohibitively high.

To solve these problems, both a wireless power transmission sheet and a communication sheet have been developed. These sheets are made with plastic MEMS switches and organic FET's as printable low-cost and large-area electronics can be embedded in furniture, walls and ceiling. These sheets detect the position of

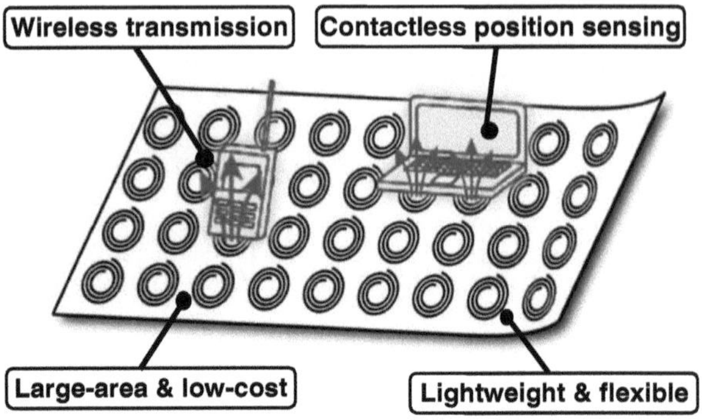

Fig. 5.4 Overview of the developed wireless power transmission sheet

the electronic device placed nearby and provide the wireless power supply and low-power wireless communication.

Sections 5.3.2.1 and 5.3.2.2 present the wireless power transmission and the communication sheet, respectively.

5.3.2.1 Wireless Power Transmission Sheet

A pictorial view of the wireless power transmission sheet [14, 15] is shown in Fig. 5.4. The developed power transmission sheet detects the position of the electronic devices on the sheet and delivers power as EM waves. The sheet is flexible, large area, and can be manufactured at low cost, because it is fabricated using a printing process on a polyimide film. The principle of the wireless power transmission is based on an array of transmission coils (TX-coil) made on the plastic sheet which are selectively driven by plastic MEMS switches and are coupled magnetically with a receiver coil (RX-coil) mounted on the power-receiving object. The typical problem of wireless power transmission using electromagnetic induction is the efficiency loss due to the displacement between the TX-coil and the RX-coil. The segmentation and selective activation of TX-coils introduced in this approach greatly reduce efficiency loss. Position detection of the receiving object is needed for the selective activation of TX-coils. This is achieved scanning through many TX-coils and measuring the antenna impedance of the TX-coil to detect if an RX-coil is placed nearby.

Figure 5.5 shows a bird's-eye photo of the manufactured wireless power transmission sheet and micrographs of coils, a MEMS switch and an OFET. Four 21 cm^2 plastic sheets are stacked to build the system. The first sheet is covered with a TX-coil array for the wireless power transmission. The second sheet has a MEMS switch array to select the TX-coils. The third sheet contains a coil array for the position detection. The fourth sheet has an OFET array to select the position detection coils. Each sheet contains an 8 $\times$ 8 array of coils, MEMS switches, and OFETs with a 1 inch pitch. MEMS switches and OFETs are used to select the TX and RX coils respectively, because they show complementary electrical properties. In MEMS switches, the on–resistance is less than 10 Ω, but the switching speed is

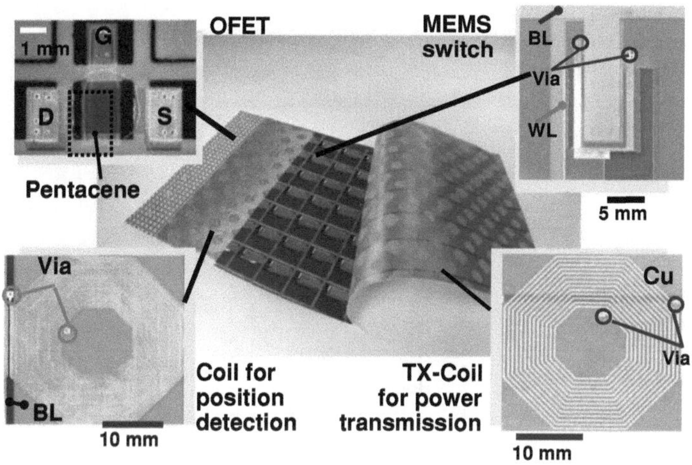

Fig. 5.5 Bird's-eye photo of the wireless power transmission sheet and micrographs of coils, MEMS switch and OFET

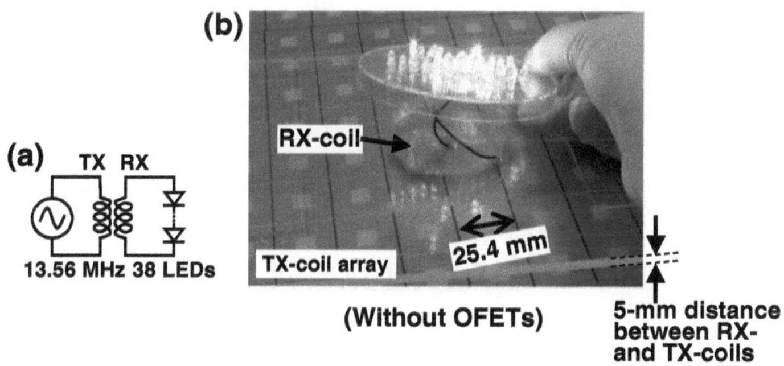

Fig. 5.6 **a** Measurement setup for the wireless power transmission to 38 LEDs. **b** Demonstration of the wireless power transmission.

1 Hz. In contrast, in OFETs, the switching speed is more than 100 Hz, but the on–resistance is more than 1 kΩ. Therefore, MEMS switches are used for the power transmission and OFETs are used for the position detection.

Figure 5.6a shows schematically the setup of an experiment to demonstrate wireless power transmission between the sheet and a string of 38 LEDs in series. Figure 5.6b shows the real-life demonstration of the wireless power transmission. The distance between RX-coil and TX-coils was 5 mm. The measured maximum power transmission efficiency was 62 % and the maximum transmitted power was 29 W [14]. These figures are limited by the parasitic resistance of the coil and could be improved using a thicker metal layer for the coil.

Figure 5.7 shows a mock-up of expected applications of the wireless power transmission sheet. The sheet embedded in the walls delivers power wirelessly to a wall-mount TV. The sheet embedded in the table delivers power wirelessly to a cell

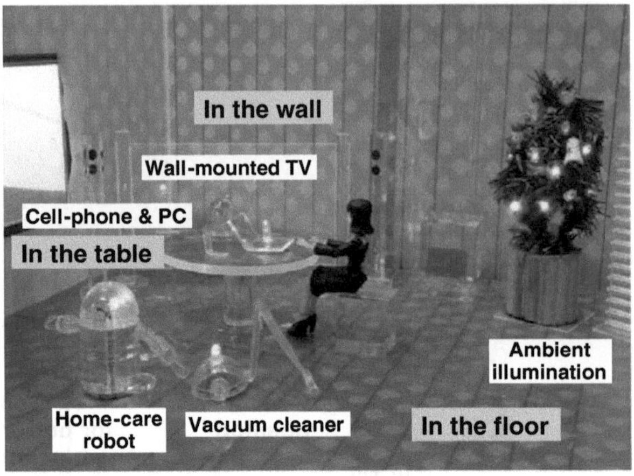

Fig. 5.7 Expected applications of the wireless power transmission sheet

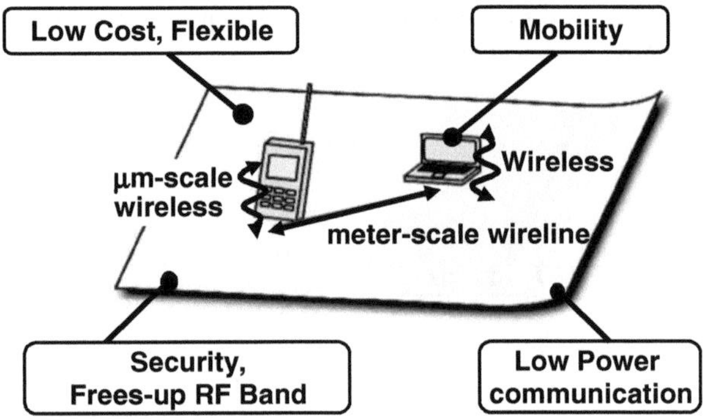

Fig. 5.8 Overview of the developed communication sheet. © 2009 IEEE. Reprinted, with permission, from [19]

phone and a laptop PC. The sheet embedded in the floor delivers power wirelessly to a moving home-care robot, a vacuum cleaner, and an ambient illumination. In this way the wireless power transmission sheet, an example of large area electronics made possible by OFET technology, will enable ubiquitous electronics.

5.3.2.2 Communication Sheet

An overview of the communication sheet that our group developed [16–19] is shown in Fig. 5.8. By combining meter-scale wireline communication and micrometer-scale wireless capacitive-coupling communication, the communication sheet combines the mobility of wireless communication and the low-power performance of wireline communication. The sheet enables multiple electronic objects scattered over tables, walls, and ceilings to communicate contactlessly with each other by establishing communication paths without cumbersome physical connections.

Figure 5.9 shows the operation principle of the communication sheet. The point-to-point communication is achieved by combining wireline communication

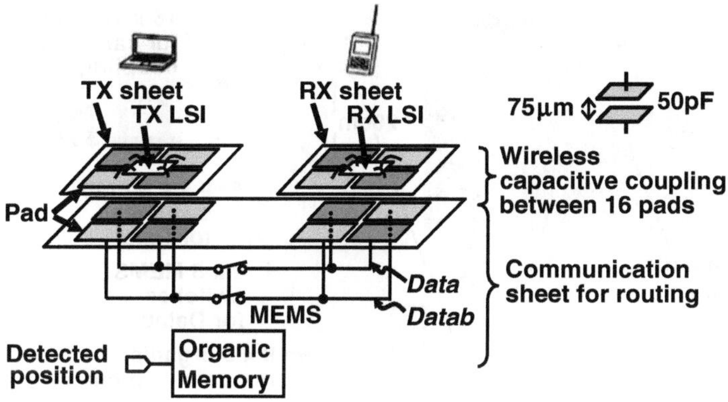

Fig. 5.9 Operation principle of the communication sheet. © 2009 IEEE. Reprinted, with permission, from [19]

within the sheet with wireless capacitive-coupling communication between the pads. The transmitter drives a pad on a TX sheet that couples to a corresponding pad and a routing line on the communication sheet, and the pad on the other end of the routing line, in turn, drives the receiver on the RX sheet. The typical pad distance for the capacitive coupling is 75 μm, which corresponds to a 50 pF capacitance. In this application, differential signaling is required because there is no common ground among the TX, communication, and RX sheets. A 90° rotation of the TX/RX sheet can be counteracted by the other two additional pads. The communication route is dynamically formed using the plastic MEMS switches, and the routing information is stored in organic nonvolatile memories [16, 18]. The "1–0" current ratio of organic nonvolatile memory exceeds 10^5, and the retention time is more than 15 days in air.

The device–level implementation of the communication sheet is shown in Fig. 5.10. It consists of a stack of five 20 × 20 cm low-cost printed sheets: a 16 × 16 pad array for capacitive coupling, two 9 × 8 MEMS switching matrices for differential signal routing, a 9 × 8 organic nonvolatile memory array for routing information storage, and an 8 × 8 position-detection coil array. The 16 × 16 capacitive-coupling pad arrays are fabricated on a polyimide film. The side length of the square pad is 9.7 mm, and the distance between each pad is 3.0 mm. Silver gate electrodes and polyimide gate dielectric layers are patterned by using inkjet printing. The MEMS switching matrix is formed by using inkjet printing and screen printing. The electrodes for electrostatic actuation of the MEMS are patterned on a 25 μm-thick polyimide membrane. Compared with an organic FET switch, a MEMS switch provides lower ON-resistance and lower parasitic capacitance, which contributes to low energy/bit communication. When 9 V is applied to the control electrodes of the MEMS switch, the resistance changes from 1 MΩ to 20 Ω. The maximum pass bandwidth of the MEMS switch is 1 kHz, while the maximum actuation frequency is 1 Hz. The position-sensing

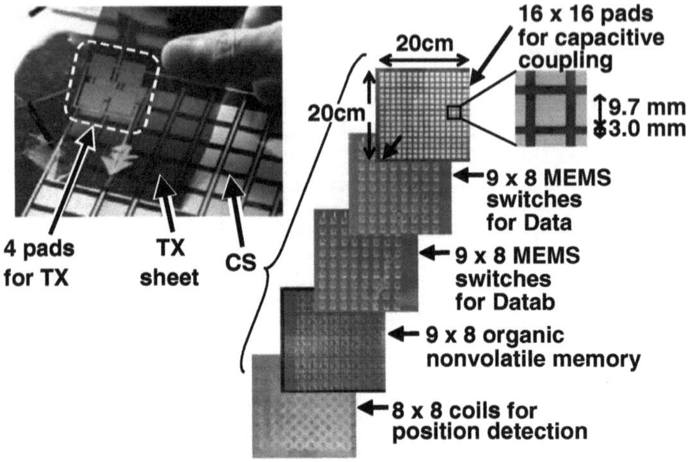

Fig. 5.10 Device structure of the communication sheet. © 2009 IEEE. Reprinted, with permission, from [19]

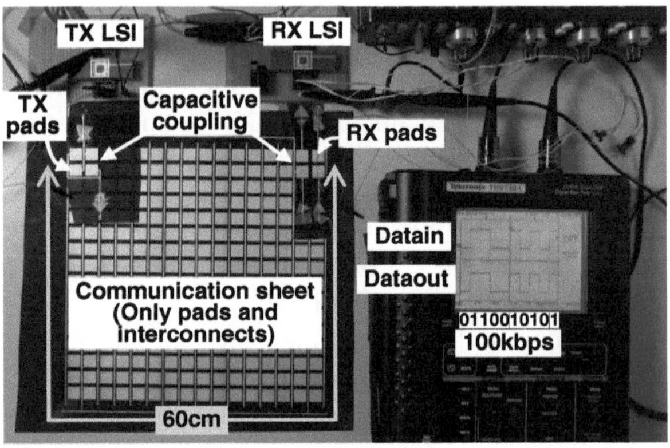

Fig. 5.11 Measurement setup of the communication sheet. A 100 kb/s communication at 10.7 μW was demonstrated at a distance of 60 cm. © 2009 IEEE. Reprinted, with permission, from [19]

coil array is manufactured by screen printing. The principle of the position-sensing sheet is similar to the one used in the power transmission sheet.

Figure 5.11 shows a measurement setup of the communication sheet. The TX/RX Si chips are bonded to the TX/RX pads on the TX/RX sheet, and the pads are capacitively coupled to the communication sheet. In this measurement, to check the feasibility of the communication through the communication sheet, the MEMS switching matrices, the organic nonvolatile memory array, and the position-detection coil array are not included, and the communication route is laid out as a 60 cm hard-wired connection. A 100 kb/s communication at 10.7 μW was demonstrated, which corresponds to 107 pJ/bit. The developed transceiver IC [17, 19] in 0.18 μm CMOS with a data-edge-signaling transmitter and a DC power-free pulse detector contributes to the low power communication.

5.4 Challenges for Future Large Area Electronics

Several challenges for the future large area electronics are itemized below.

- Improvements of organic device reliability and thus in the operational and shelf life of organic electronics are required.
- Processing should be optimized to minimize the variability of organic device parameters.
- The operation speed of organic devices should be increased without a too large impact on manufacturing cost.

The reliability of organic devices is rapidly improving by advancement in encapsulation technologies. Just as VLSI technologies have reduced the variability of silicon MOSFETs in past 40 years, the variability of organic devices will be reduced by the improvement of the manufacturing process. The miniaturization of organic FETs is a straightforward approach to increase speed; however, miniaturization normally comes at increased costs, which should be avoided to maintain the low-cost advantage of organic devices. On top of improvements in processing and devices, suitable circuit and system-level design is also a powerful tool that must be used more and more intensively to advance the state of the art of organic electronics and bring it still closer to exciting applications.

5.5 Summary

Large area electronics, which takes advantage of the flexibility and the low cost per area, is a good application for OFETs. By combining silicon VLSI and OFETs, large area electronics has the potential to expand both the application field of electronics and the total market of electronic industry. Therefore, the development of the large area electronics which sophisticatedly combines the silicon VLSI's and the OFETs is of great scientific and commercial interest.

Acknowledgments This work was partially supported by CREST/JST, the Grant-in-Aid for Scientific Research, Special Coordination Funds for Promoting and Technology, and NEDO.

References

1. Someya T, Sakurai T (2003) Integration of organic field-effect transistors and rubbery pressure sensors for artificial skin applications. In: IEEE international electron devices meeting, pp 203–206
2. Someya T, Kawaguchi H, Sakurai T (2004) Cut-and-paste organic FET customized ICs for application to artificial skin. In: IEEE international solid-state circuits conference, pp 288–289

3. Kawaguchi H, Someya T, Sekitani T, Sakurai T (2005) Cut-and-paste customization of organic FET integrated circuit and its application to electronic artificial skin. IEEE J Solid-State Circuits 40(1):177–185
4. Someya T, Iba S, Kato Y, Sekitani T, Noguchi Y, Murase Y, Kawaguchi H, Sakurai T (2004) A large-area, flexible, and lightweight sheet image scanner integrated with organic field-effect transistors and organic photodiodes. In: IEEE international electron devices meeting, pp 365–368
5. Kawaguchi H, Iba S, Kato Y, Sekitani T, Someya T, Sakurai T (2005) A sheet-type scanner based on a 3D stacked organic-transistor circuit with double word-line and double bit-line structure. In: IEEE international solid-state circuits conference, pp 580–581
6. Someya T, Kato Y, Iba S, Noguchi Y, Sekitani T, Kawaguchi H, Sakurai T (2005) Integration of organic FETs with organic photodiodes for a large area, flexible, and lightweight sheet image scanners. IEEE Trans Electron Devices 52(11):2502–2511
7. Kawaguchi H, Iba S, Kato Y, Sekitani T, Someya T, Sakurai T (2006) A 3-D-stack organic sheet-type scanner with double-wordline and double-bitline structure. IEEE Sens J 5(5):1209–1217
8. Ishida K, Masunaga N, Zhou Z, Yasufuku T, Sekitani T, Zschieschang U, Klauk H, Takamiya M, Someya T, Sakurai T (2009) A stretchable EMI measurement sheet with 8 $\times$ 8 coil array 2 V organic CMOS decoder and -70 dBm EMI detection circuits in 0.18 um CMOS. In: IEEE international solid-state circuits conference, pp 472–473
9. Ishida K, Masunaga N, Zhou Z, Yasufuku T, Sekitani T, Zschieschang U, Klauk H, Takamiya M, Someya T, Sakurai T (2010) Stretchable EMI measurement sheet with 8 $\times$ 8 coil array 2 V organic CMOS decoder, and 0.18 um silicon CMOS LSIs for electric and magnetic field detection. IEEE J Solid-State Circuits 45(1):249–259
10. Kato Y, Iba S, Sekitani T, Noguchi Y, Hizu K, Wang X, Takenoshita K, Takamatsu Y, Nakano S, Fukuda K, Nakamura K, Yamaue T, Doi M, Asaka K, Kawaguchi H, Takamiya M, Sakurai T, Someya T (2005) A flexible lightweight braille sheet display with plastic actuators driven by An organic field-effect transistor active matrix. In: IEEE international electron devices meeting, pp 105–108
11. Takamiya M, Sekitani T, Kato Y, Kawaguchi H, Someya T, Sakurai T (2006) An organic FET SRAM for braille sheet display with back gate to increase static noise margin. In: IEEE international solid-state circuits conference, pp 276–277
12. Takamiya M, Sekitani T, Kato Y, Kawaguchi H, Someya T, Sakurai T (2007) An organic FET SRAM with back gate to increase static noise margin and its application to braille sheet display. IEEE J Solid-State Circuits 42(1):93–100
13. Kato Y, Sekitani T, Takamiya M, Doi M, Asaka K, Sakurai T, Someya T (2007) Sheet-type braille displays by integrating organic field-effect transistors and polymeric actuators. IEEE Trans Electron Devices 54(2):202–209
14. Sekitani T, Takamiya M, Noguchi Y, Nakano S, Kato Y, Hizu K, Kawaguchi H, Sakurai T, Someya T (2006) A large-area flexible wireless power transmission sheet using printed plastic MEMS switches and organic field-effect transistors. In: IEEE international electron devices meeting, pp 287–290
15. Takamiya M, Sekitani T, Miyamoto Y, Noguchi Y, Kawaguchi H, Someya T, Sakurai T (2007) Design solutions for multi-object wireless power transmission sheet based on plastic switches. In: IEEE international solid-state circuits conference, pp 362–363
16. Sekitani T, Noguchi Y, Nakano S, Zaitsu K, Kato Y, Takamiya M, Sakurai T, Someya T (2007) Communication sheets using printed organic nonvolatile memories. In: IEEE international electron devices meeting, pp 221–224
17. Liu L, Takamiya M, Sekitani T, Noguchi Y, Nakano S, Zaitsu K, Kuroda T, Someya T, Sakurai T (2008) A 107 pJ/b 100 kb/s 0.18 um capacitive-coupling transceiver for printable communication sheet. In: IEEE international solid-state circuits conference, pp 292–293
18. Sekitani T, Zaitsu K, Noguchi Y, Ishibe K, Takamiya M, Sakurai T, Someya T (2009) Printed nonvolatile memory for a sheet-type communication system. IEEE Trans Electron Devices 56(5):1027–1035

19. Liu L, Takamiya M, Sekitani T, Noguchi Y, Nakano S, Zaitsu K, Kuroda T, Someya T, Sakurai T (2009) A 107 pJ/bit 100 kb/s 0.18 um capacitive-coupling transceiver with data edge signaling and DC power-free pulse detector for printable communication sheet. In: IEEE Trans Circuits SysfI Regular Papers 56(11):2511–2518
20. Magnetic field probe (CP-2S) NEC corporations
21. EMSCAN (Online) Available http://emscan.com/
22. Bode D et al (2010) Organic complementary oscillators with stage-delays below 1 μs. Appl Phys Lett 96:133307

Chapter 6
Printed RFID and Smart Objects for New High Volume Applications

Wolfgang Clemens, Jürgen Krumm and Robert Blache

Abstract Printed electronics opens up completely new application fields for electronics, where there is no electronics today. By printing conductive, semi conductive, dielectric and other functional materials in roll-to-roll processes on plastic films it is possible to realize electronic devices that are thin, flexible, low cost and available in very high volumes. Even though the overall performance of such printed electronic devices is typically lower compared to standard silicon based electronics, printed electronics could be integrated directly into packaging, to make consumer goods smart: e.g. integrated radio frequency identification (RFID) tags that transmit information in an electronic way or so called Smart Objects that enable, for example, dynamical optical elements to appear on packages, tickets or brand products. The technology of printed electronics is young and still not mature, therefore it is important to start with first products in niche markets where the advantages of printed electronics are obvious and where both the manufacturer and the user can learn about the use of such devices. To enter mass markets, some challenges still need to be faced, especially in the field of material and process optimization, but also at the system level, in order to find the best ways to integrate printed electronics with low cost goods. This chapter describes the status quo, first products and market outlook for printed electronics with a focus on printed RFID and printed Smart Objects.

Keywords Printed electronics · Organic electronics · RFID · Smart objects · Printing technology

W. Clemens (✉) · J. Krumm
PolyIC GmbH and Co. KG, Tucherstrasse 2, D-90763 Füerth, Germany
e-mail: wolfgang.clemens@polyic.com
URL: www.polyic.com

R. Blache
Materials Accelerator, C/o University of Auckland, Auckland, New Zealand

E. Cantatore (ed.), *Applications of Organic and Printed Electronics*,
Integrated Circuits and Systems, DOI: 10.1007/978-1-4614-3160-2_6,

6.1 Introduction

Printed electronics is a new field of technology. The goal is to fabricate electronic products using printing methods, which allow high volume applications at low cost. Therefore semiconductors, interconnect materials, and dielectrics should be compatible with common roll-to-roll printing processes (contrary to what happens in standard silicon electronics). Thus the two basic challenges for printed electronics are to identify a set of materials with the required electrical functionality and to formulate inks of these materials that can be printed. Organic materials—especially polymers—are found to be good candidates for printed electronics as their electrical properties can be tailored during synthesis and many of them can be processed from solution. Printed electronics devices are then realized by printing several different layers onto substrates like thin and flexible plastic films. In such way, many applications can be realized based on the platform technology of printed electronics. Examples of these applications are displays, memories, radio frequency identification (RFID) tags, sensors, photovoltaic devices, etc. Among the many possibilities of printed electronics, this chapter focuses on printed RFID and printed Smart Objects, as both of these applications use integrated circuits and are very attractive for the mass market.

In the first part of this chapter, the basic aspects of printed electronics are described, with focus on the electrical performance and fabrication. In the second part, some applications that exploit integrated circuits, like printed RFID and Smart Objects are discussed.

6.2 Printed Transistors

As in standard silicon electronics, field-effect-transistors (FETs) are the most important active components of printed electronics circuits. In contrast to standard silicon MOSFET (metal-oxide-semiconductor FET) structures, printed transistors are realized as thin film transistors (TFT) where several layer stacks are possible. Frequently used TFT structures are depicted in Fig. 6.1. These structures differ in the stacking order of the layers and the positions of the electrodes. In top-gate structures, the gate electrode is deposited as the last layer on top of all the others, whereas in bottom-gate structures, the gate is the first layer to be deposited on the substrate. In each device setup, special compatibility requirements between the different materials used in the fabrication process (layer materials, solvents) exist. For example, solvents must not harm the electrical performance of previously deposited materials.

A representative substrate material in printed electronics is polyethylene terephthalate (PET) as it is thin and flexible, compatible with printing technologies, and commercially available in large quantities at low cost. Intrinsically conductive plastics like poly (3,4-ethylenedioxythiophene)/poly(styrenesulfonate) (PEDOT/PSS), nanoparticles, or metals can be used as electrodes. It is crucial for the transistor

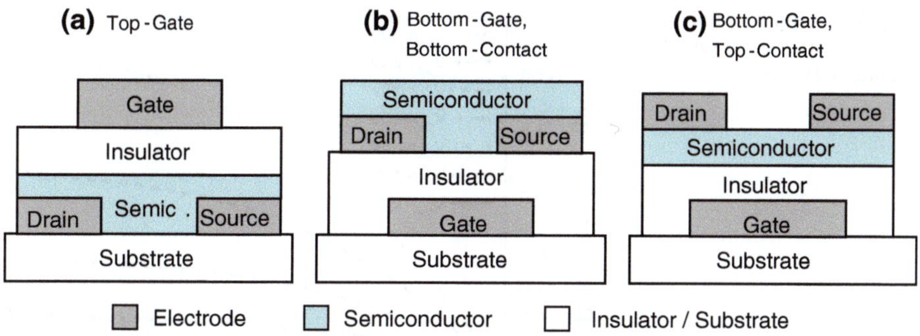

Fig. 6.1 Schematic view of frequently used structures for printed transistors (courtesy PolyIC)

performance how charge carriers are injected at the boundary between the electrodes and the semiconductor. For the semiconducting layer organic materials are widely used. In this case, the transistor is also called organic field-effect-transistor (OFET).[1] Examples of organic semiconductors are conjugated polymers like alkyl-substituted polythiophenes (e.g. poly-(3-alkylthiophene), P3AT) [1] polytriarylamine (PTAA), [2] polyfluorene derivatives (F8T2) [3] or carbon nanotubes [4].

The insulating layer separates the semiconductor layer from the gate electrode and plays the role of a dielectric. Possible candidates are any non-conductive soluble materials that form thin and homogeneous layers when processed via printing methods. Examples are common organic polymers like poly(methyl methacrylate) (PMMA) or polystyrol. An important property of the layer is the relative permittivity ε_r of the material. In a simplified approach, the transistor can be treated as a parallel-plate capacitor. Being ε_r the relative permittivity of the dielectric material, d its thickness and A the gate area, the capacitance of the transistor gate is then given as

$$C = \varepsilon_r \cdot \varepsilon_0 \cdot A/d, \tag{6.1}$$

where $\varepsilon_0 = 8.85 \times 10^{-12}$ As/Vm is the vacuum permittivity.

When switched on, OFETs normally accumulate charge carriers at the semiconductor-insulator interface. The amount of accumulated carriers and consequently the current flow in the channel between source and drain is modulated by the gate voltage and is proportional to the gate capacitance per unit area. Consequently, using a material with a higher relative permittivity or decreasing the insulating layer thickness leads to higher currents. The minimum layer thickness is determined by the maximum leakage current that can be allowed through the insulating layer. This should be significantly lower than the current between source and drain at any operation point of the transistor. Typical values for the insulator thickness are $d \leq 1$ μm. The shapes of the output characteristics (drain current versus drain-source voltage) of OFETs are comparable to that of conventional

[1] In other sections of this book OFETs are also called OTFTs.

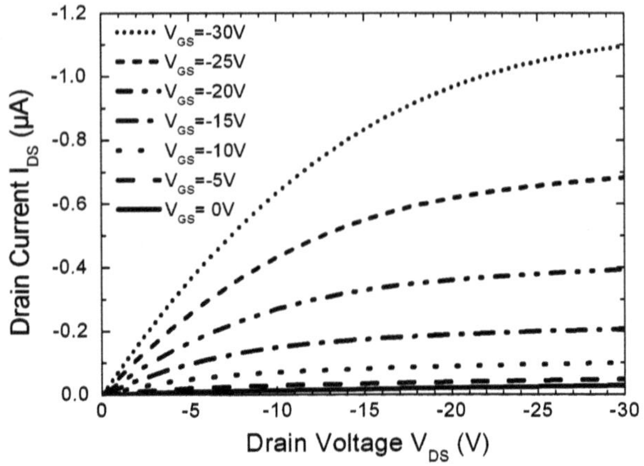

Fig. 6.2 Typical output characteristics of an OFET (courtesy PolyIC)

silicon transistors (see Fig. 6.2) but the magnitudes of the necessary voltages are typically higher than 10 V to get currents in the μA range. In principle, both n-type and p-type OFETs can be fabricated. But till recently the latter have shown far better electrical performance regarding charge-carrier mobility and stability under ambient conditions. Therefore, circuits with only p-type OFETs are still the ones mostly used in printed electronics.

6.3 Printing Technology

When soluble organic materials are used for printed electronics large-scale fabrication methods like flexography, gravure printing, screen printing, or inkjet printing, etc. are readily applicable. In Ref. [5] and in Chap. 2 of this book different printing techniques suitable for printed electronics are reviewed. Printing of polymer integrated circuits (ICs) considerably increases fabrication volume and throughput and also considerably cuts fabrication costs. High volume and reduced fabrication costs are important features of printed electronics especially when printing techniques can be used for roll-to-roll manufacturing, where the printing substrate is fed through a printing machine from one roll to another. In comparison to batch processing of IC wafers, roll-to-roll manufacturing can reach much higher throughput. As an example of roll-to-roll fabrication, gravure printing will be discussed here. In gravure printing, a flexible printing foil is continuously unwound from a roll, travels between a rotating gravure cylinder and an impression cylinder, and is finally rewound onto another roll. In the printing process, the surface of the rotating gravure cylinder is wetted by an ink or soluble polymer in a tank. The ink (semiconductor, insulator, metal-like electrode material, etc.) is scraped by a doctor blade so that only the quantity filling up the recessed part of the cylinder remains. This ink is transferred to the target position of the printing substrate while the cylinder rolls across it. Figure 6.3 shows a schematic view of the printing process described above.

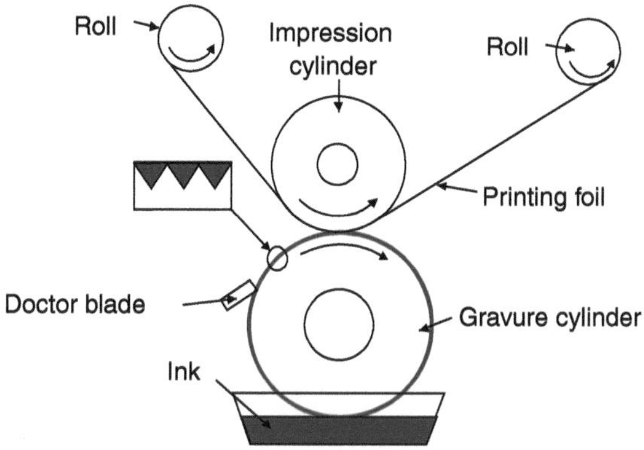

Fig. 6.3 Schematic view of gravure printing [16]

Using printing techniques also leads to some restrictions. Due to the limited printing and alignment resolution of printing machines, feasible feature sizes of printed ICs, i.e. critical dimensions, are limited to 10 μm and above [1]. Printing materials have to be chosen carefully in order to remain compatible with the respective printing technology. Printing of integrated circuits also leads to completely different requirements than in conventional printing: continuous line structures are required for the drain/source electrodes with feature sizes in the micron range. The semiconductor and the insulator have to be stacked upon each other as thin, homogeneous, and defect-free layers. Additionally, the gate structure must be deposited with high precision relative to the structures on the drain/source layer [6]. The rheological behavior, i.e. flow characteristics, of the dissolved polymers like viscosity, adhesion and coating behavior, can differ by orders of magnitude from those of conventional printing inks. Adjusting the rheological behavior by means of additives, which is done in conventional printing, is difficult, as each layer in a printed IC has a specific function. These adjustments are usually detrimental to the electrical properties of the materials.

When all challenges of printing processes are successfully met, the result can look like the printed roll shown in Fig. 6.4. The image in Fig. 6.4a shows the roll with a typical length of 2–5 km wound around a 6-inch core. The layer stack on this roll corresponds to the top gate configuration in Fig. 6.1a. The layout contains single transistors for testing purposes as well as integrated circuits. A section of the roll and a detailed photograph of the lower electrode layer with source and drain electrodes are shown in Figs. 6.4b and c. Typical properties of the underlying fabrication process are given in Table 6.1.

6.3.1 First Printed Circuits

The aim of printed electronics is to realize printed integrated circuits, which include a large number of transistors. As a simple example of such circuits, ring oscillators are often employed in assessing the feasibility of a semiconductor

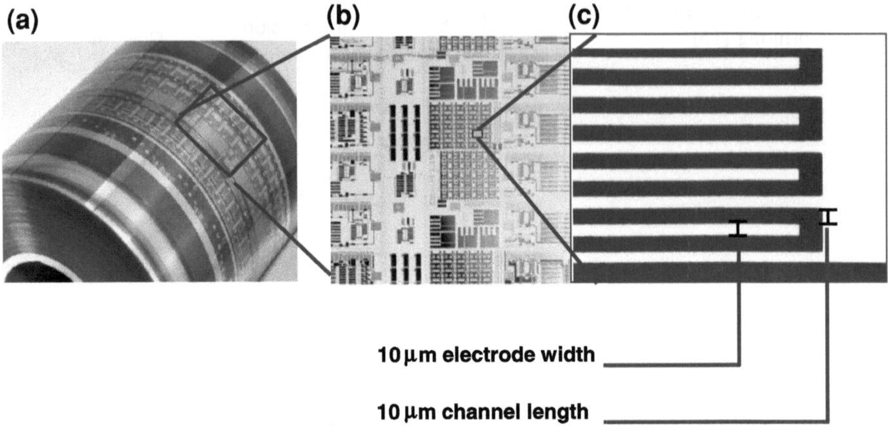

Fig. 6.4 Photograph of a roll of printed integrated circuits (courtesy PolyIC)

technology with respect to switching speeds and logic capabilities. The first printed circuit appeared in 2004 [7]. It realized a seven-stage ring oscillator fabricated by use of traditional graphic art printing techniques such as pad printing and blade coating. Roll-to-roll printed seven-stage ring oscillators were obtained with gravure printed PTAA as the semiconductor and PEDOT-PSS for the source/ drain electrodes [2]. These circuits reached oscillation frequencies of 1.2 Hz at a supply voltage of 48 V. In another publication, a seven-stage ring oscillator using F8T2 as the semiconductor was reported with an oscillation frequency of 67 Hz at 30 V [3]. The first printed RFID transponder was reported in 2007 [8]. This tag transmitted a ring oscillator signal. In 2009, a printed 15-stage ring oscillator with 107 Hz at a supply voltage of 15 V was reported by the same group [9]. Due to the restrictions of printing processes, these oscillation frequencies considerably lack behind performance figures given for organic ring oscillators fabricated with cleanroom setups, which can be faster than 600 kHz [10].

More complex printed circuits were also demonstrated. For example, Jung and colleagues report on a gravure-printed 4-bit code generator for an RFID transponder with carbon nanotubes as the semiconductor [4]. A printed circuit, which integrates all components necessary for an RFID transponder circuit transmitting a Manchester-encoded 4-bit sequence, was shown in 2009 [9]. The circuit consists of about 200 devices and contains an organic rectifier, a ring oscillator as clock generator as well as a protocol generator including a counter and a ROM. More details of the circuit are given in Ref. [11].

Besides printed p-only organic electronics, recent developments include printable organic complementary transistors (see Sect. 6.5), printable ZnO transistors [12] or printed silicon transistors [13]. These developments will lead to increased performance if the materials can be processed with a roll-to-roll approach on polyester films.

Table 6.1 Typical parameters for a roll-to-roll printing process in printed electronics (courtesy PolyIC)

Property	Value
Typical web length	2–5 km
Typical web speed	>30 m/min
Substrate thickness	50 μm
Electrode layer thickness	20–80 nm
Dielectric/semiconductor layer thickness	50–500 nm
Weight of roll	50–200 kg
Area per roll	1000–3000 m^2
Critical dimension	10 μm
OFET structure	Top-gate

6.4 Important Performance Factors for Printed Electronics

Application of printing methods in printed electronics comes at the cost of performance limitations. Some of the issues in printed electronic circuits are summarized in this section.

6.4.1 Transistor Speed

The speed of a transistor in an integrated circuit can be coarsely characterized by the g_m/C_{in} ratio, with g_m being the transconductance and C_{in} the input capacitance of the device [14]. This quantity can be approximated as $g_m/C_{in} = \mu \cdot (V_G - V_T)/L^2$, where μ is the charge carrier mobility of the transistor, L is the channel length, V_G is the gate voltage, and V_T the threshold voltage of the device. The cut-off frequency f_c of a single transistor can then be estimated as

$$f_c = \frac{g_m}{2\pi C_{in}} = \frac{\mu \cdot (V_G - V_T)}{2\pi L^2} \tag{6.2}$$

Assuming e.g. $\mu = 0.02\ \text{cm}^2/\text{Vs}$, $L = 10\ \mu\text{m}$, $V_G - V_T = 10$ V, a cut-off frequency of $f_c \sim 32$ kHz can be expected. This analysis has some shortcomings as it does not include peculiarities of printed transistors like large overlap and parasitic capacitances, V_{GS}-dependent charge carrier mobilities, etc., which lead to significantly lower speeds for real printed circuits than one would expect from Eq. (6.2). This formula, however, serves as a first-order approximation to the attainable circuit speed and shows that increasing mobility and/or decreasing channel length are the key factors for implementing faster circuits. The switching frequency and the current that printed circuits can drive are significantly limited compared to standard electronics, and the size of a printed IC is significantly larger than the size of a standard IC.

6.4.2 Yield

Circuit yield is affected by the real-estate usage of the layout and the defect densities obtained in the fabrication process. For example, in Ref. [1] results on the characterization of 2818 printed transistors on a 3 km-long roll were reported. There, 98.5 % of the devices were identified to lie within specified limits. Figure 6.5 shows the histogram of the drain currents of these devices as well as the spatial distribution over the roll (equivalent to wafer maps commonly plotted for silicon wafers). Both the drain-source and gate-source voltage were equal to −20 V in all measurements.

In the following, circuit yield Y will be analyzed using the basic approximation

$$Y = p^N \tag{6.3}$$

Here, p is the probability for a single device to be within specification and N is the number of devices in a circuit. A device yield of 98.5 % leads to an approximate circuit yield of 22 % for a circuit containing 100 devices. In this simplified analysis, only the device yield was considered: defect contributions relating e.g. to interconnect lines were neglected. An example of a circuit with a complexity of about 100 devices is a 3-stage counter consisting of conventional master–slave RS flip-flops (8 two-input NORs with 3 devices per NOR = 24 devices for each stage) with an 11-stage ring oscillator realized by inverters (11 inverters with 2 devices for each = 22 devices) and some buffers.

A special issue is parametric yield, where the variations of the parameters of individual devices will lead to a certain percentage of fabricated circuits not conforming to the specifications but being functional. The parametric yield of roll-to-roll fabricated circuits has yet to be studied in more detail in the literature. It can, however, be anticipated that sustaining a defined level of parameter variation along kilometer-long foils is difficult. Consequently, a considerable amount of global variation can be expected. This is less important for comparatively robust digital circuits but will considerably complicate the design of analog circuit blocks like analog-to-digital converters or sensor amplifiers.

6.4.3 Threshold Voltage Engineering

The threshold voltage of the transistors in a process technology is an important parameter in circuit design. Currently, it is difficult to deliberately tune the threshold voltage to the requirements of printed electronics. For silicon transistors

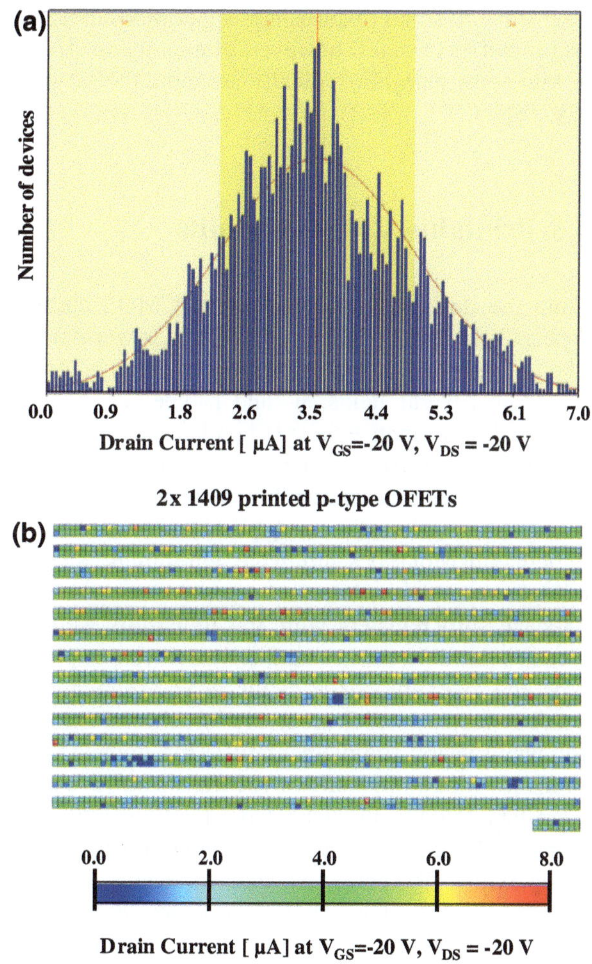

Fig. 6.5 Distribution of the drain current of 2818 printed transistors from one single roll: **a** histogram and **b** spatial distribution over roll (courtesy PolyIC)

the tuning is done by doping, but organic semiconductors are intrinsic materials without intentional doping. Material development here still has to find methods to deliberately adjust the threshold voltage of printed transistors. As a consequence, designers have presently to cope with this limitation of the printed transistor technology. In p-type transistors, normally-off behavior would be often desirable. Normally-off or enhancement mode means that the transistor is completely switched off at a gate-source voltage $V_{GS} = 0$ V. Printed p-type transistors often exhibit weakly normally-on or depletion mode behavior. A normally-on transistor is not switched off at a gate-source voltage $V_{GS} = 0$ V, which is due to a positive threshold voltage $V_T > 0$ V for p-type transistors. Already in the early days of silicon n-type digital logic, enhancement and depletion mode transistors were combined to design logic gates with high noise margins, i.e. logic gates that are insensitive to small disturbances superimposed on the input signal. In state-of-the-art printed electronics, however, often only one transistor type is used. This results in low noise margins, typically around 2.0 V at supply voltages of $|V_{DD}| = 20$ V [15, 16].

6.5 Printable CMOS Circuits

From the designer's point of view, CMOS circuits would be highly favorable especially in terms of noise margins and power-delay product. But printable CMOS-like processes are only in a beginning phase and come along with additional challenges in several fields: n- and p-type semiconductor materials with sufficient electrical performance have to be identified, possibly the source and drain contacts have to be modified in order to efficiently inject both positive and negative charge carriers, dielectrics compatible with both semiconductors have to be used, and finally high resolution printing processes have to be developed that allow integration of both transistor types within the same circuit. The first organic CMOS-type circuits of higher complexity based on organic semiconductors appeared in 2001 [17], when 48-stage shift registers with 864 transistors were presented. These circuits were clean-room fabricated using fluorinated copper phthalocyanine ($F_{16}CuPc$) as the n-type and α-sexithiophene as the p-type semiconductor. Both semiconductors are small molecules, which are rarely soluble and thus not well suited for printing. In the following years, complementary D flip-flops [18] and a 4-Bit CMOS RFID transponder with a complexity of 168 transistors [19] were presented. These circuits still were fabricated using clean-room facilities although soluble materials were used. Printed CMOS-type circuits presented in the literature currently lack behind these complexities. Recently, Ng and colleagues [20] reported on an ink-jet printed CMOS inverter using poly (2,5-bis(3-tetradecyl-thiophene-2-yl) thieno(3,2-b)thiophene) (PBTTT) as the p-type and ActiveInk N1400™ from Polyera as the n-type semiconductor. Various printing methods have been reported for all-printed polymer CMOS inverters [21]. Lately, printed 5-stage ring oscillators with CMOS inverters were presented with an oscillation frequency of approx. 11 Hz at a supply voltage of 40 V [6]. A derivative of PTAA was used as the p-type semiconductor and a derivative of perylene diimide as the n-type semiconductor.

6.6 Advantages and Limitations of Printed Electronics

To discuss both the advantages and the limitations of printed electronics it is worth summarizing briefly what printed electronics compared to standard silicon electronics is all about.

As mentioned before, printed electronics means that conductive, semiconductive and dielectric materials are used as "electronic" inks in printing processes, where they are applied onto substrates like thin and flexible plastic films. Typical materials are organic polymers like poly(alkyl)thiophene, which are soluble in organic solvents. There are many different printing processes that can be used to print electronics. This enables fabrication of electronic devices that are thin, flexible and low cost in very high volume.

In contrast, standard electronics is typically based on integrated circuits made of silicon ICs, which are produced in batch processes and clean rooms. Continuously increased circuit complexity has made possible all the electronic devices of our daily life, like computers or mobile phones. Silicon ICs are our reliable daily companions in many areas, but limits are where very thin and flexible electronics at low cost for single use is demanded. Therefore, today, one hardly finds single use or disposable electronic devices, mainly because of cost reasons.

Printed electronics devices are as thin and flexible as the e.g. 50 μm thin polyester films (see Table 6.1) on which the devices are printed. The printed multilayer stack building the transistor typically consists of noncrystalline or amorphous layers with a total thickness around or less than 1 μm, which maintains the flexibility.

What about the costs? Even though the raw materials, especially the organic semiconductors are high tech chemicals which are still relatively expensive, only a small amount of these materials is needed. In the case of silicon, the whole wafer must be made of highly pure and costly semiconductor, in printed electronics the substrate is a low cost material and the valuable semiconductor can be printed e.g. by additive processes where there is no material loss (contrary to the subtractive processes that are typically used in standard electronics).

The electrical performance limitations of printed electronics circuits are mainly due to two reasons: the performance of the used electronic materials and the resolution of the printing process. These limitations are important to know when thinking about applications of printed ICs. But printed electronics enables new ways to bring electronics to everyday applications, which today are out of reach for standard electronics and probably will remain such in the foreseeable future. Here printed electronics can gain special interest due to the possibility for integration and low cost at high volumes that it offers.

Therefore, printed electronics is a new platform technology that enables many different applications. As we have seen so far the printed electronics devices are different from the standard ones and should be used when the application can strongly benefit from printing technology. In general, printed electronics suits large area, thin, flexible, high volume and low cost applications.

6.7 Main Application Fields of Printed Electronics

Many devices implemented with standard electronics, such as displays, memories, RFID-tags, sensors, photovoltaic cells, etc. can also be realized using printed electronics. Printed electronics serves as a platform technology as discussed in the

Fig. 6.6 Working principle of RFID. *Left side* RFID tag, *right side* RFID reader (courtesy PolyIC)

roadmap of the organic electronic association [22] and in Chap. 1 of this book. Among the many possibilities of printed electronics, the remaining part of this chapter focuses on printed RFID and printed Smart Objects. Both of these applications use integrated circuits and are very attractive for the mass market. Integrated circuits consist of transistors, diodes, capacitors and resistors that are combined to realize suitable functions. In the case of RFID such an IC is combined with an antenna, in the case of Smart Objects it is combined with other elements like sensors or displays.

6.8 Printed RFIDs

An RFID system consists of a reader and at least one RFID tag as shown in Fig. 6.6. The reader emanates radio-frequency waves at a given frequency. The RFID tags are tuned to this frequency. The RFID tag consists of a transponder chip (an IC), mounted on top of an antenna. The RF field of the reader powers up the IC of the tag. The IC responds with the stored information by modulating the RF field. This enables RF communication over a certain distance without the need for line of sight.

Printed RFIDs are a highly interesting application for printed electronics for many reasons. RFIDs represent a huge market in applications like identification, logistics or automation. With the so called electronic product code (EPC), it is planned to replace the widely used optical barcode for consumer products by RFID tags. This will enable a significant improvement of the whole supply chain of goods, from production to transport, through warehouses and retailers, to the end customer. The vision here is to have a reader in the shop basket of the supermarket which directly identifies all goods that are put into it. This leads to many advantages for the customer, like a more convenient buying or intelligent shelves

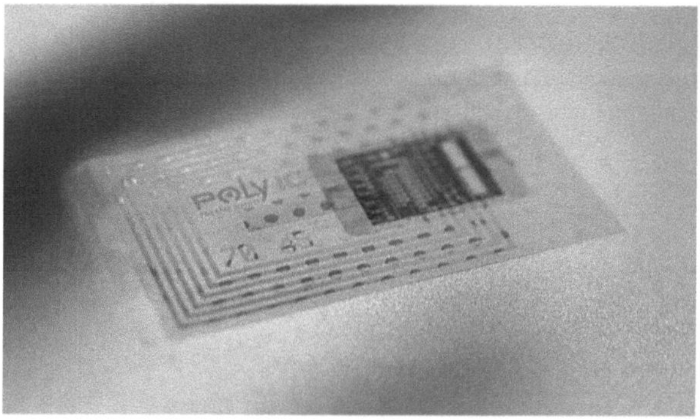

Fig. 6.7 Printed RFID tag consisting of a printed integrated circuit and an antenna (courtesy PolyIC)

that are never "out of stock". Standard silicon-based RFID tags could implement these applications but are limited by relatively high costs and are difficult to integrate into the packaging material. Here, printed RFIDs like the one shown in Fig. 6.7 provide crucial advantages: they can be significantly cheaper, and they are thin and flexible. The principal component of printed RFID is an IC realized with printed transistors. This circuit is like a plastic sticker which is mounted on top of the RFID antenna. The size of the printed IC is considerably larger than a comparable silicon IC, but due to the size of the antenna this is not a problem. The transmission needs to be adjusted to international standards; here the so called high frequency range (HF) at 13.56 MHz is the most widely used frequency for printed RFID. In the future printed RFID might also be available in the ultra high frequency (UHF) range, close to 1 GHz.

6.9 Printed Smart Objects

Printed Smart Objects are devices which combine different functions, either by direct printing on a substrate or by a hybrid integration of different printed components on a common substrate. Combining different functional devices by printing is easier than with the standard manufacturing processes used in conventional electronics. As a long term vision, smart systems will combine energy sources like a battery or photovoltaic cells with flexible display elements, memories and sensors, all functionally linked together by printed ICs. As one example in this field, Smart Objects that combine RFID functions with display elements are of major interest. Here the advantages of RFID tags—no on-board energy source, no need for line of sight—are combined with the possibility to view the information stored within the tag on an integrated display. The schematic setup of such a smart object is presented in Fig. 6.8.

Printed Smart Objects like radio frequency driven displays can be used in many applications, like brand protection, ticketing, marketing, or identification. As an

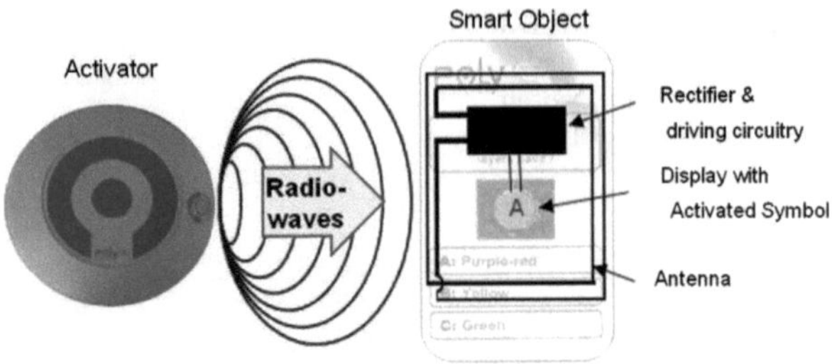

Fig. 6.8 Principle of a printed Smart Object that combines RFID with optical features (courtesy PolyIC)

example, a display element for an electronic quiz card is shown in Fig. 6.9. The depicted Smart Object combines the authentication of a quiz card with an interactive feature. The answer is only displayed for a genuine card held close to a reader.

A next step for the Smart Objects will be the integration of sensors with printed circuits for the measurement of environmental parameters like temperature and humidity, chemical or biomedical quantities.

6.10 Market Entry for Printed RFID and Smart Objects

Printed electronics provides considerable prospects for new products and markets. However, the main challenge is to find the right market entry, i.e. positioning novel devices in application fields where they are superior to existing solutions or where other technologies are difficult to employ.

In this section the possible market entry strategies for printed RFID and Smart Objects are discussed. On one hand there are established markets for these objects, especially in the field of RFID. On the other hand, printed devices allow completely new applications.

In the case of printed RFID, the established identification markets can only be entered when the printed tags are compatible to existing standards. As this will need some time from a technical viewpoint, the focus for market entry lies in those fields where proprietary standards can be used and the advantages like flexibility and low cost are more important than sheer electronic performance.

One very interesting field for printed RFID is brand protection. According to a report of the European Commission [23] the number of goods, suspected of infringing intellectual property rights, detained in 2009 within the countries of the European Union, was 118 million articles. This results in huge economic losses worldwide. Besides the losses, counterfeits are coupled to risks like quality issues, absence of certification, or threats to health in the case of medical products. Therefore, especially in the field of medical devices or pharmaceuticals, an electronic integrated brand protection feature based on printed RFID is of great interest.

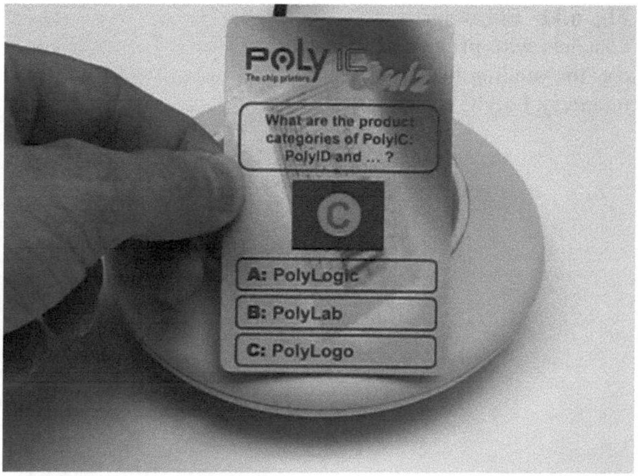

Fig. 6.9 Example of a printed Smart Object as electronic quiz card with radio frequency activated display element (courtesy PolyIC)

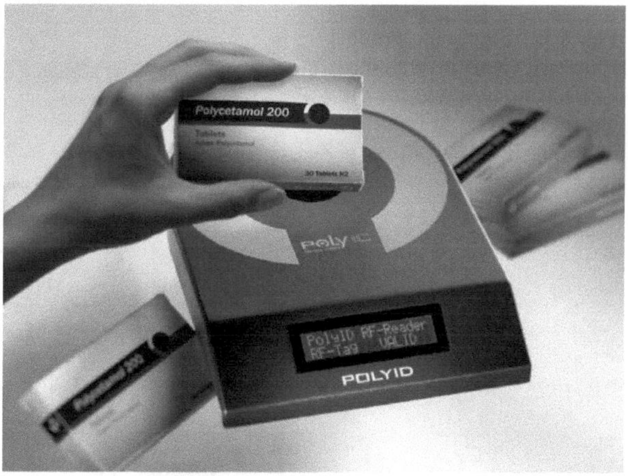

Fig. 6.10 Electronic brand protection with printed RFID in pharmaceutical applications (courtesy PolyIC)

In addition to the widely used optical brand protection features like holograms, printed RFID enables an automatic and fast readout with clear information given at the display of the reader, see Fig. 6.10. Other areas are also of interest for brand protection or authentication, like reliable identification of genuine consumables.

Electronic vouchers are another relatively new field for RFID. Single use printed RFID tags are suitable for a number of applications e.g. as vouchers for a self service coffee machine, as shown in Fig. 6.11.

Further applications for printed RFID or Smart Objects are in electronic ticketing, automation, or logistics.

In the long term, the focus for printed RFID lies in the realization of the electronic product code—EPC tag that will substitute the optical barcode in logistics and on consumer products (Fig. 6.12).

Fig. 6.11 Electronic vouchers with printed RFID, e.g. for vending machines (courtesy PolyIC)

Fig. 6.12 Electronic product codes (EPC) with printed RFID as a substitute for the optical bar code in retail (courtesy PolyIC)

This application will only be possible at item level, where the opportunities offered by printed electronics in terms of cost and integration possibilities in combination with the ability to produce large quantities of devices are a distinctive advantage.

In the case of printed Smart Objects, the market entry will probably take place with applications that have an event character, as there is hardly an established market for this. It is important to demonstrate the maturity of the products in first applications, which can be niche markets. Examples are special marketing applications where companies can demonstrate that they use innovative technologies. After first reference products have successfully been placed in niches, higher volume markets for sensor applications, security, brand protection, information labels, raffle tickets, etc. can be targeted.

6.11 Main Challenges and Next Steps for Printed Electronics on its Way Toward Mass Markets

The major challenge for printed electronics is to reach a level of maturity which enables market entry in different fields of applications. At present, there are first printed electronic products on the market but not yet in high volume applications. The challenges are manifold as this is a completely new platform technology based on new materials and new processes. There are many possible candidates to realize so called "killer applications" for printed electronics, and we discussed in some detail two examples: printed RFID and printed Smart Objects. It is expected that printed electronics will not compete with standard silicon-based electronics but will complement it, defining completely new fields of application.

The performance of materials for printed electronics is progressing rapidly, especially in the fields of semiconductors and conductors; the printing processes are also developing fast, as many companies and institutes are currently working in these fields. It is, however, a major challenge to find the optimum set of materials and processes for the specific applications.

The next steps for printed electronics will be to find the right area for market entry for each product. These areas must be open for new technologies and printed electronics will enable new features in flexibility and cost that cannot be enabled by standard electronics. After entering the niche markets with proprietary standards and overcoming the main technology challenges, mass markets will be the next step. In mass markets, global standardization and the general availability of common infrastructure are of great importance. One vision of printed RFID is the electronic product code on the yoghurt cup. When this will be realized is still matter of debate, but there are many other high volume markets waiting for electronic "intelligence" which can be realized by printed electronics.

References

1. Fix W (2010) R2R printed electronics. In: Presentation at Lope-C conference. Frankfurt, Germany, 1–2 June 2010
2. Zielke D, Hübler A, Hahn U, Brandt N, Bartzsch M, Fügmann U (2005) Polymer-based organic field-effect transistor using offset printed source/drain structures. App Phys Lett 87:123508
3. Klink G, Hammerl E, Drost A, Hemmetzberger D, Bock K (2005) Reel-to-reel fabrication of integrated circuits based on soluble polymer semiconductor. In: Presentation at IEEE Polytronic 2005
4. Jung M, Jung K, Lim SY, Lee K, Kim DA, Kim J, Tour KM, Cho G (2009) All R2R printable 4-bits digital signal processor for printed RFID Using SWNT-TFTs", ICFPE, Jeju Island, Korea
5. M enard E, Meitl MA, Sun Y, Park JU, Shir DJL, Nam YS, Jeon S, Rogers JA (2007) Micro- and nanopatterning techniques for organic electronics and optoelectronic systems. Chem Rev 107(4):1117–1160

6. Jacob S, Gwoziecki R, Verilhac JM, Benwadih M, Seiler AL, Bory C, Bablet J, Heitzmann M, Tallal J, Altazin S, Boudinet D, Courant Y, Fischer V, Mohamed F, Frères P, Sicard G, Chartier I, Coppard R, Serbutoviez C (2010) Full printed organic CMOS circuits for large area electronics. In: Presentation at Lope-C conference. Frankfurt, Germany, 1–2 June 2010
7. Knobloch A, Manuelli A, Bernds A, Clemens W (2004) Fully printed integrated circuits from solution processable polymers. J Appl Phy 96(4):2286–2291
8. Fix W (2007) Polymer based 13 MHz RFID transponders. In: Organic electronics conference 2007, Frankfurt, Germany, 24–26 Sept 2007
9. Fix W (2009) R2R printed electronics. In: Presentation at Lope-C conference. Frankfurt, Germany, 23–25 June 2009
10. Fix W (2004) Fast and stable integrated polymer circuits. At OSC conference. Cambridge, UK, 27–28 Sept 2004
11. Krumm J, Clemens W (2010) Printed electronics—first circuits, products, and roadmap. AACD Workshop, Graz, 23–25 Mar 2010
12. Volkman SK, Mattis BA, Molesa SE, Lee JB, de la Fuente Vornbrock A, Bakhishev T, Subramanian V (2004) A novel transparent air-stable printable n-type semiconductor technology using ZnO nanoparticles. In: International Electron Device Technical Digest, 2004, pp. 769–773
13. Kovio, Inc. (2011) Kovio achieves printed electronics milestone with world's first all-printed high-performance silicon thin-film transistor, press release. http://www.kovio.com/pdf/kovio_pr_11_13_07.pdf. Accessed 2011
14. Crawford RH (1967) MOSFET in circuit design. McGraw-Hill, New York
15. De Vusser S, Genoe J, Heremans P (2006) Influence of transistor parameters on the noise margin of organic digital circuits. IEEE Trans Electron Devices 53(4):601–610
16. Krumm J (2008) Circuit analysis methodology for organic transistor. Dissertation, University of Erlangen-Nuremberg, Germany
17. Crone BK, Dodabalapur A, Sarpeshkar R, Filias RW, Lin YY, Bao Z, O'Neill JH, Li W, Katz E (2001) Design and fabrication of organic complementary circuits. J Appl Phys 89(9): 5125–5132
18. Yoo B, Madgavkar A, Jones BA, Nadkarni S, Facchetti A, Dimmler K, Wasielewski MR, Marks TJ, Dodabalapu A (2006) Organic complementary D flip-flops enabled by perylene diimides and pentacene. IEEE Electron Device Lett 27(9):737–739
19. Blache R, Krumm J, Fix W (2009) Organic CMOS Circuits for RFID Applications. In: IEEE International Solid State Circuits Conference, pp. 208–209
20. Ng T, Sambandan S, Lujan RA, Arias AC, Newman C, Yan H, Facchetti A (2009) Electrical stability of inkjet-patterned organic complementary inverters measured in ambient conditions. App Phys Lett 94:233307
21. Yan H, Chen Z, Zheng Y, Newman C, Quinn JR, Dötz F, Kastler M, Facchetti A (2009) A high-mobility electron-transporting polymer for printed transistors. Nature 457:679–686
22. Organic Electronics Association (2009) Organic and printed electronics, 3rd edn. VDMA Verlag, Frankfurt, Germany. (www.oe-a.org)
23. European Commission (2009) Report on EU customs enforcement of intellectual property rights—results at the EU border 2009. http://ec.europa.eu/taxation_customs/resources/documents/customs/customs_controls/counterfeit_piracy/statistics/statistics_2009.pdf. Accessed 2011

Chapter 7
Organic RFID Tags

Kris Myny, Soeren Steudel, Peter Vicca, Steve Smout, Monique J. Beenhakkers, Nick A. J. M. van Aerle, François Furthner, Bas van der Putten, Ashutosh K. Tripathi, Gerwin H. Gelinck, Jan Genoe, Wim Dehaene and Paul Heremans

Abstract Organic RFID tags are increasingly gaining credibility as a possible low-cost barcode replacement for product identification. This will only happen if organic RFID tags can operate in the frequency range defined by well-accepted EPC standards. This chapter evaluates the performance of existing organic RFID demonstrators and confirms that, based on lab scale demonstration of organic RFID tags, performance comparable to the EPC standards can be obtained. Moreover, the integration of sensors with the tags will enable added functionality and applications beyond pure identification.

Keywords Organic RFID tag · Flexible RFID tag · Flexible electronics · Organic circuits · EPC · Pentacene · Organic rectifier · Double half-wave rectifier · Sensor integration · RFID roadmap

K. Myny (✉) · S. Steudel · P. Vicca · S. Smout · J. Genoe · W. Dehaene · P. Heremans
imec, Kapeldreef 75, 3001 Leuven, Belgium
e-mail: myny@imec.be

K. Myny · W. Dehaene · P. Heremans
Katholieke Universiteit Leuven, Leuven, Belgium

K. Myny · J. Genoe
Katholieke Hogeschool Limburg, Diepenbeek, Belgium

M. J. Beenhakkers · N. A.J.M. van Aerle
Polymer Vision, Eindhoven, The Netherlands

F. Furthner · B. van der Putten · A. K. Tripathi · G. H. Gelinck
TNO Science and Industry, Eindhoven, The Netherlands

N. A.J.M. van Aerle
now at ASML, Veldhoven, The Netherlands

E. Cantatore (ed.), *Applications of Organic and Printed Electronics*, Integrated Circuits and Systems, DOI: 10.1007/978-1-4614-3160-2_7,

7.1 Introduction

Due to the low-temperature budget and the potential for high-throughput of their manufacturing technology, thin-film organic semiconductor transistors lend themselves to develop applications where mechanical flexibility and thin plastic substrates are important, like backplanes for flexible displays and smart objects integrating electronics, communication and even sensors. A core technology indispensible for all these applications is the development of organic circuits: we illustrate here the state of the art of this technology with special emphasis on the design of transponder chips for RFID applications.

In this chapter, we first outline the longer-term goals and requirements to be set for thin-film and plastic transponder chips. They are derived from the aspiration to be compatible with existing Si standards, in particular the Electronic Product Code standards. Then, we review the current state of the art in organic TFT chips, concentrating on the digital function (i.e. the code generator) of an RFID tag. We also show that all functions of a transponder chip besides the code generator, i.e. the rectifier and the modulator, can be made with present plastic transistor technology, so that complete transponder chips and operational tags can be demonstrated. In the last part, we show a first example of integration of a sensor with plastic RFID tags, which paves the way to smart labels.

7.2 Organic RFID Roadmap

RFID is an important technology in logistics, retail, automation, anti-counterfeit protection, identification and several other fields. It allows for transmission of data (an identification code in the memory of the transponder chip) via radio waves from a transponder to a reader without the need for line of sight. In standard, silicon based electronics, this is a large market, segmented in several areas of applications. Several radio communication frequencies are used, spanning from tens of kHz to the low GHz range.

The goal of organic RFID tags is clearly the market of low-cost tags and labels. Therefore, we consider here only passive tags, which merely comprise an antenna and a transponder chip and need no battery (see Fig. 7.1). The antenna of the tag is designed to resonate at the frequency of the electromagnetic wave emitted by the reader antenna. The transponder chip has the following three functions: (1) a rectifier on the transponder chip produces a DC voltage from the antenna voltage, with which the rest of the chip is powered up; (2) the bit sequencer reads out data from a non-volatile memory as a sequence of bits; (3) a modulator varies the impedance of the antenna in synchronous with the bit sequence. The reader can decode the data by sensing the modulation of the absorbed power of the reader antenna.

Item-level tagging—where RFID replaces identification means like barcodes on each retail item—is potentially a very large market for low-cost passive organic

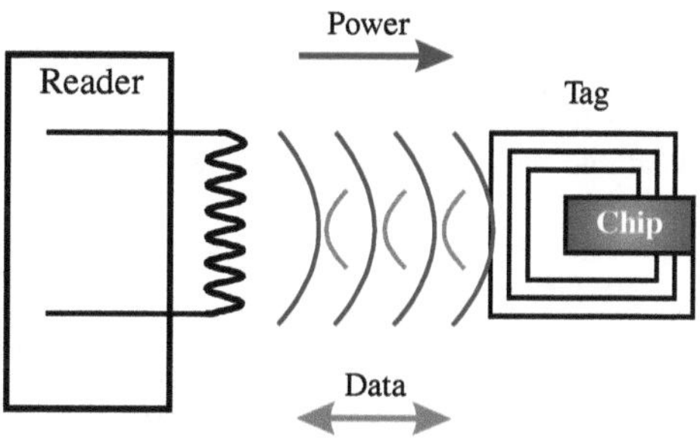

Fig. 7.1 Schematic overview of a passive RFID system

RFID tags. In that respect, a development goal for organic RFID technology is to become compatible with standards which have already been defined for that application. Indeed, this would ensure compatibility with installed infrastructure and future interoperability.

EPC Global is an organization entrusted by industry, established to create and support the Electronic Product Code (EPC). This is a global convention for immediate, automatic, and accurate identification of any single item in the supply chain of any company, in any industry, anywhere in the world. The tag is only a token to access distributed and replicated data bases. These conventions rely on the ISO norms, especially the ISO 18000 family. EPC is widely used already today e.g. on pallet level logistics. The next step is to use EPC tags at the package level and the longer term target will be its use for individual items or, in other words, for item-level tagging.

One of the major limits of the EPC tags on their way towards item-level tagging is the price of the single silicon transponder chip on the tag. The costs of the tags (or "inlays" according to a widespread naming) are dominated by the price of the silicon chip, followed by the cost of the antenna and of the assembly. For low-cost products, the price of the silicon transponder chip is significantly too high. Bringing this cost down is very challenging, because it is proportional to the chip size, which cannot be decreased at will because of the minimum size needed for economic handling and die-attachment on the inlay.

Printed thin-film electronics could create a break-through in this respect. Lower cost per chip could result from the use of fewer process steps, less energy in the production (resulting from the lower process temperature), the use of low-cost substrates (which saves packaging costs) and higher processing speeds (resulting in a lower depreciation cost of manufacturing equipment per chip).

Furthermore, the availability of a thin-film circuit directly on a plastic substrate confers superior mechanical properties to the assembled tag. Conventionally, the lamination of the RFID "inlay" in cardboard, paper or thin plastic results in an end-product with uneven topology, due to inlay thickness variations caused by the presence of the rigid silicon chip—and this limits the possibilities of roll-to-roll handling and lamination to products. Thinned-down silicon chips exist, but they are more expensive and often more fragile which limits the throughput of die attachment into inlays. In contrast, thin-film chips are thin and flexible, which solves both the handling and the yield issues.

Table 7.1 Main technology-related specs of EPC Gen 1 and Gen 2

Summary of EPC specs	EPC ISM band class 1	EPC class 1 generation 2
Carrier frequency (f_c)	13.56 MHz	860–960 MHz
Baud rate interrogator to label	26.48 kb/s (f_c/512)	26.7–128 kb/s
Baud rate label to interrogator	52.969 kb/s (f_c/256)	Baseband modulated: 40–640 kb/s Subcarrier modulated: 5–320 kb/s
Number of bits	64 or 96 +16 bit redundancy +24 bit destroy ("kill") code	Same, plus access password of 16 bit
Energy	Passive (no battery)	Passive (no battery)
Anti-collision	Label must be selectable in a group of related labels (interrogator talks first)	Label must be selectable in a group of related labels (interrogator talks first)
Destroy	Individually destroyable	Individually destroyable
Max. power of interrogator	Europe: 42 dB μA/m at 10 m	Europe: 42 dB μA/m at 10 m
Memory	Programmable	Programmable
Corresponding ISO/IEC norm	ISO/IEC 18000-3	ISO/IEC 18000-1, ISO/IEC 18000-6

Since item-level tagging should be the goal of organic thin-film RFID tags, and since EPC global proposes standards applicable to the item-level tagging application, it is important to verify that the EPC standards could be reached with organic thin-film technology. Within EPC, several classes and generations can be distinguished. In Table 7.1, the main technology-related specifications of EPC systems at 13.56 MHz (HF) and UHF are listed for passive tags.

At UHF communication frequency, the antennas can be printed, can make use of less-conductive metal tracks, and are smaller compared to at HF. That implies that UHF allows for lower-cost antennas and is therefore the best solution for the lowest cost tags. However, the rectifier at UHF requires diodes with multi-GHz cut-off frequency, which have not been shown in thin-film technologies (organic or based on other semiconductors) so far.

Therefore, demonstrations so far concern HF tags. Their antenna is a resonant circuit comprising a coil with very low resistance and a capacitor. Several demonstrations of code generators, transponder chips and HF tags have been shown, and will be discussed in the next paragraphs.

Even for HF tags, some of the specifications of EPC have not been demonstrated yet with plastic chip technologies. A key one is the demand of bi-directional communication. To avoid collision of data at the reader when multiple tags are present simultaneously in the field of the reader, the protocol of data transfer from tag to reader specifies that the reader should talk first and give instruction to the tags one by one to transmit their code. This asks for the possibility to demodulate, at the tag side, information transmitted by the reader. Furthermore, the data clock should be derived from the HF field, and not be generated as an independent and asynchronous clock on the tag. These features are beyond the technological capabilities of plastic chip technology today, because they demand fast transistors and circuit blocks operating at the frequency of the HF field, i.e. at 13.56 MHz. Achieving these features is a goal for future research.

A further point of consideration is the nature of the non-volatile memory containing the code. Implementations so far were hard-wired memories, generated at the site of production of the tag. Clearly, WORM (write-once read-many) memory could already be a great an improvement over this, since the programming could be done at the site of use. A plastic WORM technology that could be envisaged in this respect is shown in [1] using fuses of poly(3,4-ethylenedioxythiophene) poly(styrenesulfonate) (PEDOT:PSS). Ultimately, an electrically re-programmable plastic memory compatible with and embeddable in thin-film plastic circuitry would be the most versatile solution. The required number of bits is modest, from 64 for the simplest tags, to kbit for more elaborate versions. Several possible memory elements that could serve this scope, like ferroelectric diodes and transistors, have been demonstrated already.

Finally, it should be noted that thin-film technologies are also ideal to make (large-area) sensors or actuators that could be integrated with RF tags. Furthermore, also display effects such as electrophoretic, electrochromic or even OLED can be integrated with thin-film and organic transistors. The thin-film circuit technology can thus act as versatile platform for smart labels that sense, display and transmit information.

7.3 State of the Art of Organic RFID Tags

In recent years, several research groups have studied and published research results on organic RFID systems. In 2007, Cantatore et al. published a capacitively-coupled RFID system where a 64-bit code was read out at a base carrier frequency of 125 kHz [2]. The 64-bit code generator was fully functional at a 30 V supply voltage. In that pioneering work, code generators with a limited number of bits (up to 6 bit) could be read out using a base carrier frequency of 13.56 MHz by a capacitive antenna. Ullmann et al. demonstrated a 64-bit tag working at a bit rate exceeding 100 b/s, readout by inductive coupling at a base carrier frequency of 13.56 MHz [3]. In this section, we review recent advances in both the digital transponder chip and the analog front-end of organic RFID tags, and demonstrate that organic electronics can result in a tag with a realistic code size, bit rate and reading distance at reasonable and allowed field strength [4–8]. Finally, we demonstrate the possibility for sensor integration in RFID systems.

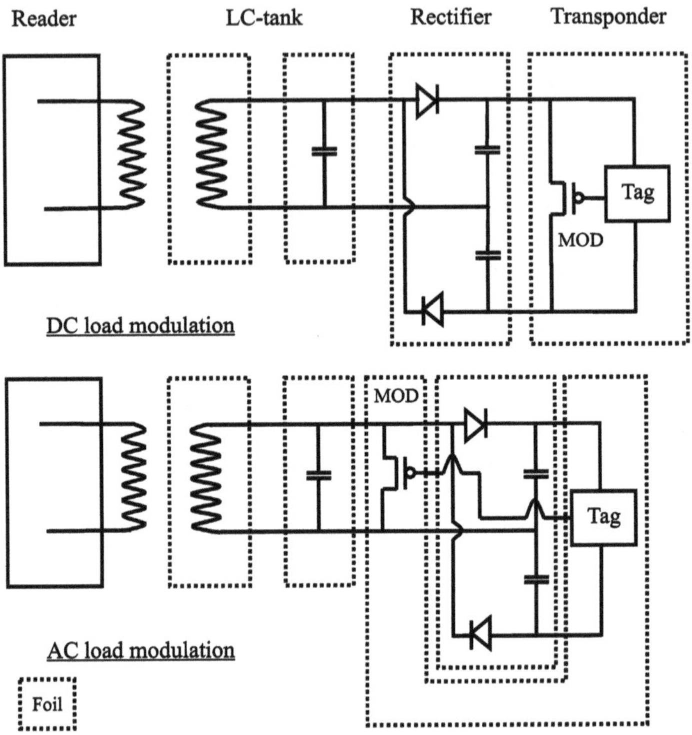

Fig. 7.2 Inductively-coupled organic RFID tags using DC (*top*) and AC (*bottom*) load modulation[1]

7.3.1 Building Blocks for a Passive RFID Tag

The basic schematic of a passive, organic RFID tag realized by our group is depicted in Fig. 7.2. The RFID tag consists of 4 different modules: the antenna coil, the HF-capacitor, the rectifier and the transponder chip with an integrated load modulator.

The coil and the HF-capacitor form an LC tank resonating at the HF resonance frequency of 13.56 MHz, which provides the energy for the organic rectifier with an AC voltage at 13.56 MHz. The rectifier generates the DC supply voltage for the organic transponder chip (or code generator), which drives the modulation transistor between the on- and off-state with the code sequence to be transmitted. Load modulation can be obtained in two different modes, depending on the position of the load modulation transistor in the RFID circuit, as shown in the two panels of Fig. 7.2. AC load modulation, whereby the modulation transistor is placed in front of the rectifier, sets demanding requirements to the organic thin-film transistor (OTFT),

[1] © 2009 Elsevier. Reprinted, with permission, from [5].

since it has to be able to operate at HF frequency. This is not obvious, as a consequence of the limited charge carrier mobility of OTFTs, which ranges nowadays 0.1–1 cm^2/Vs for typical organic semiconductors like pentacene. Therefore, load modulation at the output of the rectifier (DC load modulation) is preferred in organic RFID tags. In the latter mode, the OTFT does not require to operate at HF frequency. All organic RFID tags discussed in this chapter operate in DC load modulation mode.

7.3.2 Organic Transponder Chips

The organic transponder chips in this work comprise a load modulator and a code generator. In this subsection, we will focus on the design and measurements of the code generators. The main purpose of these code generators is to read out multiple bits of code from hardwired memory and send them to the load modulator. More complex code generators have also been demonstrated, which include Manchester encoding and a basic anti-collision protocol [5–7]. Here two code generators are shown, having 64 and 8 bits of hardwired memory available, but fabricated in two different technologies [5–8].

The basic schematic of a 64-bit code generator is depicted in Fig. 7.3. When powered, a clock signal is generated by a 19-stage ring oscillator. This clock signal is used to clock the output register, the 3-bit binary counter and the 8-bit line select. The 8-bit line select has an internal 3-bit binary counter and a 3-to-8 decoder. This block selects a row of 8 bits in the code. The 3-bit binary counter drives the 8:1 multiplexer, selecting a column of 8 bits in the code matrix. The data bit at the crossing of the active row and column is transported via the multiplexer to the output register, which sends this bit on the rising edge of the clock to the modulation transistor. The 3 bits of the 3-bit binary counter are also used in the 8-bit line select block for selecting a new row after all 8 bits in a row are transmitted.

The organic TFT technology that is used for this transponder chip was developed by Polymer Vision for the commercialization in rollable active matrix displays and is described elsewhere [9]. The organic insulator layers and the p-type pentacene semiconductor are processed from solution. The transistors have a typical channel length of 5 μm and an average saturation mobility of 0.15 cm^2/Vs.

The 64-bit code generator comprises only 414 OTFTs. At 14 V supply voltage the 64-bit transponder foil generates the correct code at a data rate of 752 b/s, as depicted in Fig. 7.3. This foil has been used for the realization of an integrated, 64-bit, organic RFID tag (see Sect. 7.3.4 in this chapter).

The data rate of the previous transponder chip is limited to values below 1 kb/s, however in order to fulfil the data rate requirements of an EPC-based RFID ransponder chip, 52.969 kb/s for HF (Table 7.1), the design of an 8-bit transponder chip has been elaborated on an organic technology that enables high data rates [8]. The critical factor to address the circuit speed is the current drive of the transistors, determined by the carrier mobility, the specific capacitance of the gate dielectric, and the inverse of the channel length. Vapor-deposited pentacene [8, 10] has a

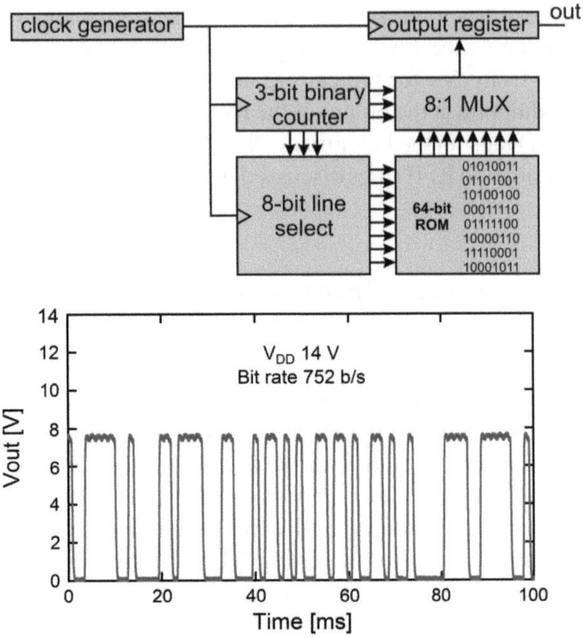

Fig. 7.3 (*Top*) schematic overview of the digital logic part of the 64-bit transponder chip. (*Bottom*) output signal of the 64-bit transponder chip at a supply voltage of 14 V (see footnote 1)

mobility, in our bottom-contact devices, exceeding 0.5 cm^2/Vs. As gate dielectric, we use sputtered Al_2O_3, with a specific capacitance of 70 nF/cm^2, treated by a self-assembled monolayer of trichloro(phenethyl)silane. That, in turn, allows for some downscaling of the transistor channel length, in our case to 2 μm. In this work, we varied the channel lengths (L) of the circuits between 20 and 2 μm. We also limited the parasitic gate-source and gate-drain overlap capacitances by decreasing the width of the finger-shaped source and drain contacts (that overlap the gate) from 5 to 2 μm. 19-stage ring oscillators, which are the clock generators in the 8-bit code generators, have been fabricated and measured. Figure 7.4 depicts the measured stage delay as a function of the supply voltage for ring oscillators with varying channel lengths of the transistors, from 20 to 2 μm and with varying gate-overlap of the transistor-fingers ranging from 5 to 2 μm. Stage delays below 1 μs, and as low as 400 ns, are shown at V_{DD} as low as 10 V. The effect of decreasing the overlap capacitance is also shown in Fig. 7.4 for the circuits having a channel length of 2 and 5 μm: shrinking the overlap from 5 to 2 μm improves the stage delay by a factor of 1.5–2.

Furthermore, an 8-bit code generator has been designed and fabricated for transistors having channel lengths and finger widths of 2 μm. Figure 7.4 depicts the signal obtained from the 8-bit code generator when powered with 18 V, yielding a data rate of 50 kb/s that is compatible with EPC specifications.

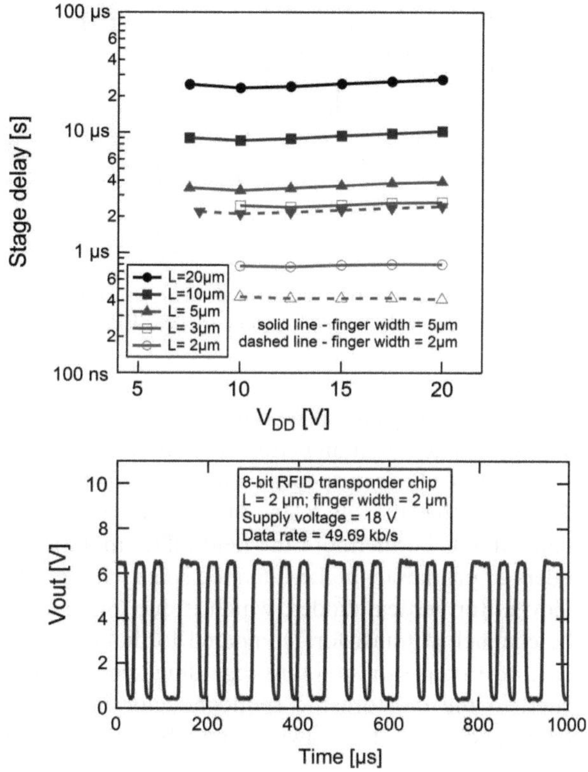

Fig. 7.4 (*Top*) overview of stage delays versus supply voltage, measured on 19-stage ring oscillators with varying channel lengths and source/drain finger sizes. (*Bottom*) measured signal of the 8-bit RFID transponder chip having transistors with channel length and source-drain finger size of 2 µm. © 2010 Elsevier. Reprinted, with permission, from [8]

7.3.3 HF and UHF Rectification

In the general introduction, it has already been pointed out that in order to be a competitive and viable product, the organic RFID tag should work at the standard base-carrier-frequency of 13.56 MHz [11] (high frequency—HF) or 869/915 MHz (ultra high frequency—UHF) which is challenging because of the intrinsically low charge carrier mobility in organic semiconductors. The rectification stage in its simplest realization consists of a diode in series with a load capacitance to which the other stages of the circuit are connected in parallel as a load. The rectifying stage is, by far, the most critical part of the RFID tag for what speed is concerned, because the rectifying element needs to be able to charge the capacitance of the rectifier in only a fraction of the carrier frequency period. The circuitry of the logic operates at a much lower frequency, which depends on the required data rate. The gate of the load modulator transistor is also driven at this lower data rate, whereas the source-drain voltage follows the base-carrier frequency for AC load modulation. Even in this kind of modulation the specifications for the load modulator transistor are less stringent than for the rectifying element, due to the much larger voltage over the element and the availability of the full period to modulate.

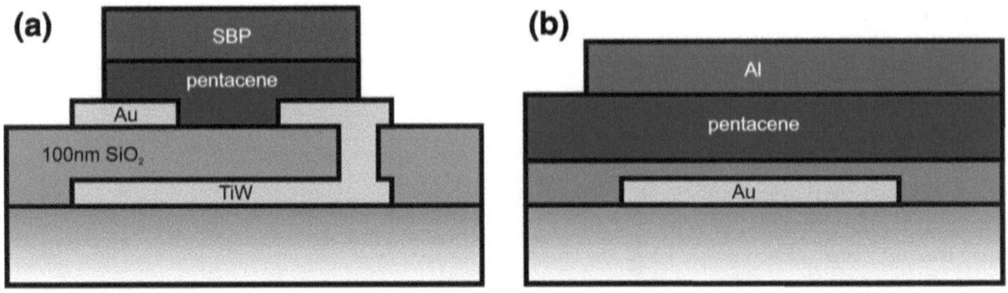

Fig. 7.5 Cross-section of experimental diode structures: (**a**) organic field effect transistor with shorted drain-gate, (**b**) vertical organic diode. SBP stands for semiconductor barrier protection

Moreover, only a partial modulation of the antenna load current is already sufficient to enable communication with the reader.

It has already been suggested to use an OTFT with shorted gate-drain as a diode [12], which has the advantage of easy integration in the same process steps as the logic circuit. Furthermore the field-effect mobility shown for OTFTs is generally orders of magnitude higher than the space-charge-limited current (SCLC) mobility relevant for vertical diodes, which is due to the higher charge density achieved in the channel of transistors compared to diode structures [13]. One obvious drawback of a transistor-structure regarding speed is the long transistor channel compared to the thin organic semiconductor layer that can be realized in vertical Schottky diodes. This drawback may become even more important in the future as the process of fabricating transistors shifts from laboratory-scale high-resolution photolithography processes, that allow channel-lengths down to 1 μm, to high-volume production, which use cheap printing processes but enable only lower resolution (typically >10 μm).

7.3.3.1 Realization of Discrete Vertical Diodes and Their DC Characteristics

To evaluate the intrinsic frequency limits of organic diodes, we have fabricated pentacene transistors on glass with 100 nm sputtered SiO_2 as gate-dielectric as described in detail elsewhere [14]. A diode-connected transistor is a transistor in which the gate and drain nodes are shorted. For this set of experiments, transistors with different channel length and width are made on the sample (Fig. 7.5a). Vertical diodes of different sizes using highly-purified, vacuum-deposited pentacene [15] (Fig. 7.5b) have been fabricated too. Pentacene was chosen because, at the state of the art, it can be considered the baseline material for maximum mobility in organic layers. The injection electrode is a lithographically patterned line of 20 nm Au, which may or not be covered by 30 nm PEDOT/PSS as a hole-injection layer. The semiconducting layer is 300 nm of pentacene deposited at a temperature of 20 °C with a deposition rate of ~3 Å/s. The top Al-electrode is deposited through a shadow mask.

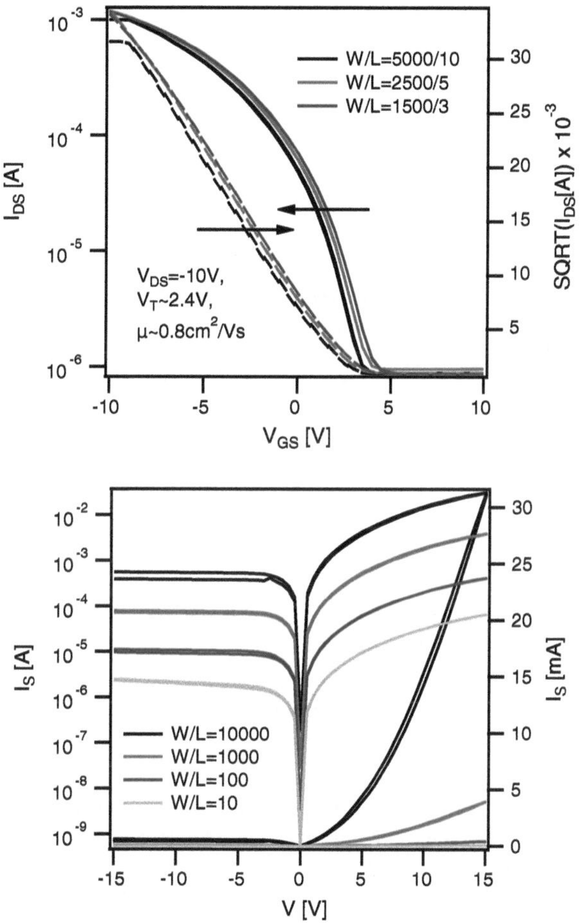

Fig. 7.6 Transistor based diode: (**a**) transfer curve of the transistor, (**b**) current–voltage characteristics of the diode. Reprinted with permission from [24]. Copyright 2006 American Institute of Physics

The chosen geometries were designed for the two different devices to be able to operate them in a similar voltage range. The structures sustain a forward bias of >15 V, a reverse bias of >25 V and achieve a high current while keeping the parasitic capacitance low. The chosen voltage range derives from the requirement to drive organic circuits, which operate mostly in a range from 10 to 15 V.

Figure 7.6a shows the transfer curves ($I_D - V_G$) in saturation of transistors made with our process. The transistors with a W/L of 5000/10, 2500/5, 1500/3 have a mobility of $\mu \sim 0.8$ cm^2/Vs and a threshold voltage $V_T \sim 2.4$ V. Figure 7.6b shows the I–V characteristics of transistor with shorted gate-drain and various W/L. The diode-connected transistors exhibit a good rectifying behaviour with an I_{on}/I_{off}—ratio of $\sim$100. The positive threshold voltage causes the relatively high off current.

The I–V characteristics of the vertical diode structures are depicted in Fig. 7.7a, both for samples with a PEDOT/PSS injection layer (full line) and an UV-ozone cleaned Au injection layer (long dashes).

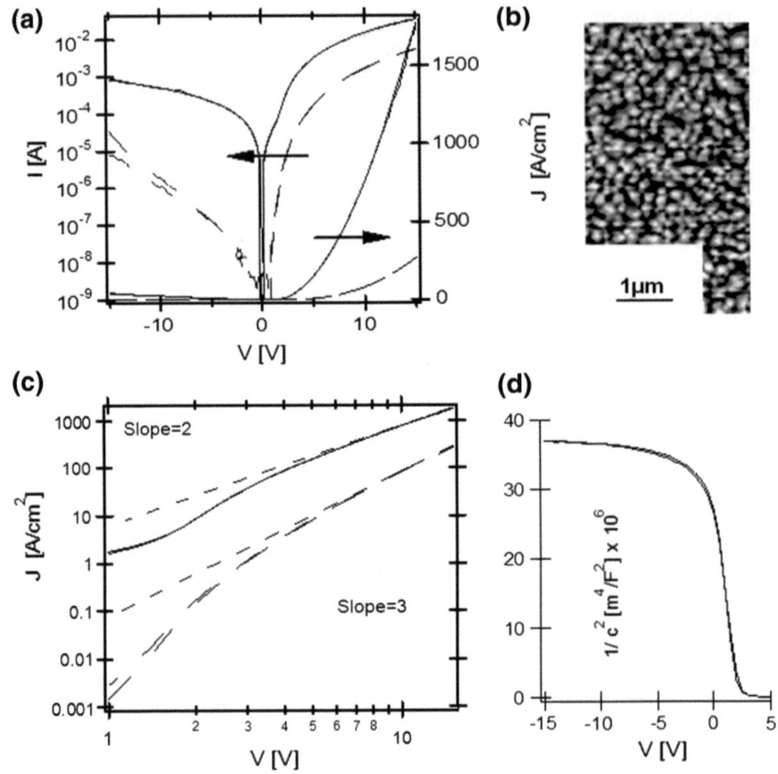

Fig. 7.7 Vertical pentacenepentacene diode: (**a**) Current–voltage characteristics (A ~ 170 × 10 μm). The continuous *line* represents samples with a PEDOT/PSS injection layer whereas the *long dashes* represent samples with a UV-ozone cleaned Au injection layer. (**b**) Atomic force microscopic image of polycrystalline pentacene, (**c**) log–log plot of IV curve under (**a**), (**d**) capacitance–voltage plot. Reprinted with permission from [24]. Copyright 2006 American Institute of Physics

The samples with a PEDOT/PSS injection layer reach a current density exceeding 2,000 A/cm^2, roughly 5 times higher as compared to the UV-ozone cleaned Au injection layer sample. However, the reverse current is significantly better in the diode without PEDOT/PSS. The high reverse current in the sample with PEDOT/PSS is mostly due to the lateral current between the probe needles. Separately fabricated samples with high-conductive PEDOT/PSS showed a much higher reverse current compared to samples with low-conductive PEDOT/PSS. Scratching the pentacene and PEDOT/PSS around the diode area improved the off current. Patterning the organic semiconductor as well as the PEDOT/PSS might thus significantly improve the rectification ratio. Measuring these samples in air decreased the maximum current by more than one order of magnitude, which might be due to a reduced hole-mobility due to humidity [16] or to the oxidation of the Al cathode. On a Si/SiO_2 substrate the diode sustains a forward bias of 19 V and a reverse bias larger than 30 V. Using a glass substrate the breakdown voltage under forward bias is reduced to 7 V, which can be traced back to the low heat conductivity of the glass. This lower breakdown voltage on glass is only an issue for the DC measurement of the IV curve. Measuring under HF conditions (MHz), the diode on glass can work in the same voltage range as a comparable diode on a Si/SiO_2 substrate. From the double-logarithmic plot shown in Fig. 7.7c, it can be seen that the current–voltage (IV) characteristics of the sample without PEDOT/PSS injection layer (long dashes) has a slope significantly larger than 2, which implies injection-limited charge transport [17]. In contrast, the sample with PEDOT/PSS exhibits a slope of 2 over a voltage range of nearly one order of magnitude, which is an indication of trap-free space-charge-limited transport (SCLC). Therefore it is possible to extract the hole-mobility with help of Child's law [17] from of the I–V characteristics of the device with PEDOT/PSS. The thickness of the pentacene layer after deposition, as measured with a profilometer is 300 nm. However, it must be kept in mind that the pentacene layer is polycrystalline as depicted in Fig. 7.7b, and therefore the relevant thickness is the equivalent electrical thickness. We extracted this thickness of the polycrystalline pentacene layer in the device from high frequency ($f = 100$ kHz) capacitor-voltage measurements in reverse bias. This measurement results in a value of ~160 nm (Fig. 7.7d). Based on this thickness, Child's law applied to the I–V characteristics of Fig. 7.7c gives a mobility of 0.15 cm^2/Vs. No significant difference of the value for the depletion capacitance has been observed between the samples with PEDOT/PSS and without PEDOT/PSS.

7.3.3.2 High Frequency Measurements of Organic Diodes in Rectifier Configuration

We performed high-frequency measurement of the discrete transistor diode and vertical diode in a rectifier configuration. All measurements were done in a N2-purged glove box to prevent degradation. At frequencies over 1 MHz, the cable of the measurement setup distorts the measurement signal. Therefore we designed the measurement setup depicted in Fig. 7.8.

The incoming sinusoidal signal is terminated at the probe head with a 50Ω termination to reduce reflections. The rectified signal is measured with an oscilloscope set to 1 MΩ input impedance (functioning as load) with a 1nF load capacitance in parallel, acting as smoothing capacitor.

The measurement of a transistor-diode on glass for an AC amplitude of 20V can be seen in Fig. 7.9. The 3 dB-frequency is around 10 MHz. Even so, the measurement data demonstrates that a passive RFID-tag at the HF-base carrier frequency of 13.56 MHz should be feasible.

In the next step, we measured our vertical pentacene diode described before with the same measurement setup but with a lower AC amplitude of 15 V. At 13.56 MHz, the rectified DC signal is ~10 V. Up to a frequency of 433 MHz, the rectified DC signal remains equal even though the AC fluctuation on top of the rectified DC signal increases, as can be seen in Fig. 7.10. Above the 500 MHz,

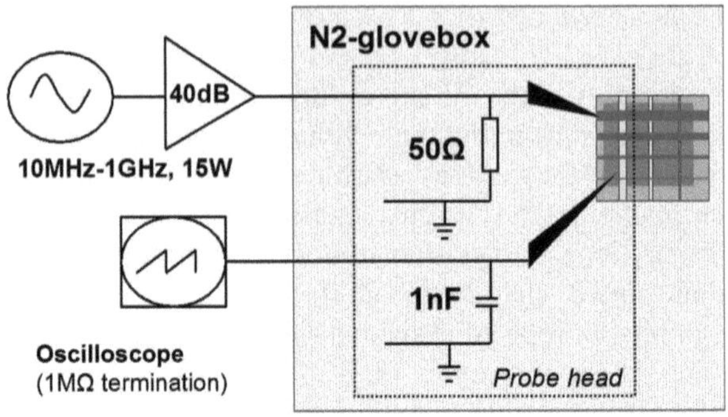

Fig. 7.8 Setup for measuring the frequency dependent rectification of discrete diodes

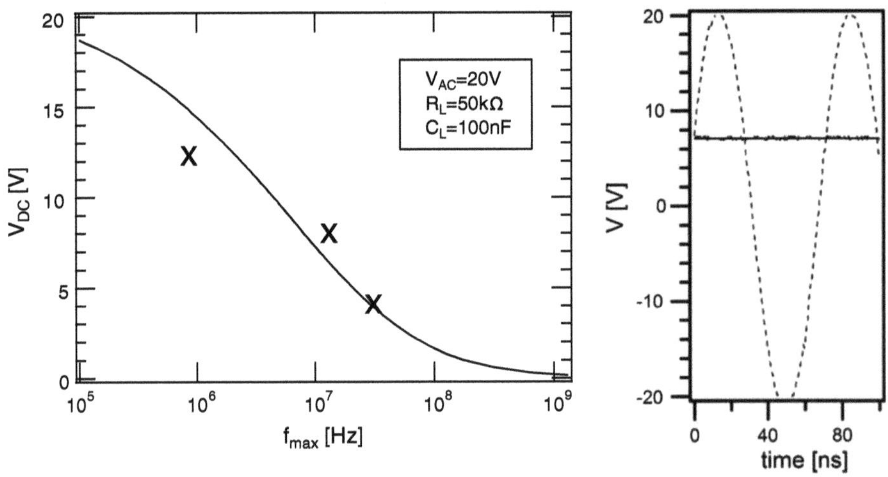

Fig. 7.9 *Left*: calculated (*lines*) and measured (*crosses*) frequency dependence of the rectified DC voltage for a rectifier with a transistor diode. The full line is calculated based on a transistor with shorted drain-gate and the following values are assumed: $\mu = 0.8$ cm^2/Vs, tox = 100 nm, $V_T = 2.4$ V, L = 3 μm, $L_{overlap} = 6$ μm and VAC = 20 V; *right*: input AC signal (*dotted line*) and rectified DC signal (*solid line*) at 13.56 MHz. Reprinted with permission from [24]. Copyright 2006 American Institute of Physics

the measurement setup proved unreliable even when using a Si-diode as reference. The important AC component remaining in the rectified voltage leads to strong reflections at the 1 MΩ input impedance of the oscilloscope.

Subsequently, we realized a diode on foil with the smoothing-capacitance (400pF) integrated in the same process-flow as the diode (see Fig. 7.11).

The electrical characteristic of the integrated vertical diode on PEN is similar to the discrete crossbar diode on glass as can be seen in Fig. 7.11c. The difference lies in the much higher reverse leakage of the integrated diode and in the much lower breakdown voltage in forward bias because of the lower heat capacity of PEN vs. glass.

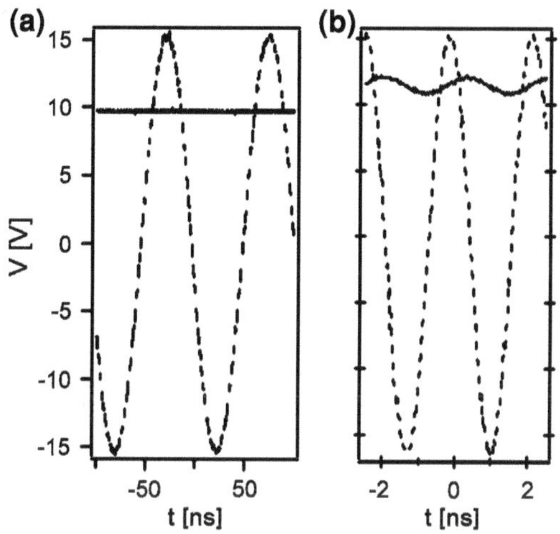

Fig. 7.10 AC input signal and rectified DC signal for a crossbar diode on glass at (**a**) at 13.56 MHz; (**b**) at 433 MHz

This rectifier foil has been placed to be measured in a socket on a PCB (printed circuit board) that contains a 50Ω input-termination and a 1 MΩ load (Fig. 7.12).

In Fig. 7.13, the rectified DC voltage using a crossbar diode as well as the integrated rectifier on foil are plotted at different frequencies for an AC input signal of 15 V amplitude. It can be seen that the integrated rectifier delivers a DC voltage of 4.5 V at a base carrier frequency of 869 MHz. The stronger drop of the rectified DC output of the integrated rectifier on PEN is probably caused the significantly higher reverse leakage of the diode [4].

It can be seen that frequencies up to 13.56 MHz are intrinsically possible for both diode configurations, i.e. the diode-connected transistor configuration and the vertical diode configuration. The maximum frequency achievable is more than two orders of magnitude higher for the vertical diode. A vertical diode may even allow operating in the UHF frequency range. For the transistor-diode there is already a significant loss of the rectified DC voltage at 13.56 MHz.

High frequency performance as well as area consumption are in the transistor based diode strongly dependent on the possible resolution and alignment accuracy of the process. In contrast, the main challenge to get sufficient performance from a vertical diode is in the high purity and low trap density of the organic semiconductor that are required to achieve SCLC mobility in the same order of magnitude as the field effect mobility. Resolution and alignment accuracy play a less significant role using a vertical diode.

7.3.4 Integrated Organic RFID Tag

The complete tag is realized by properly interconnecting the contacts of the four foils containing the four building blocks (Fig. 7.2), which we achieved in an experimental set-up where we plug the individual foils into sockets. Alternatively, we have also

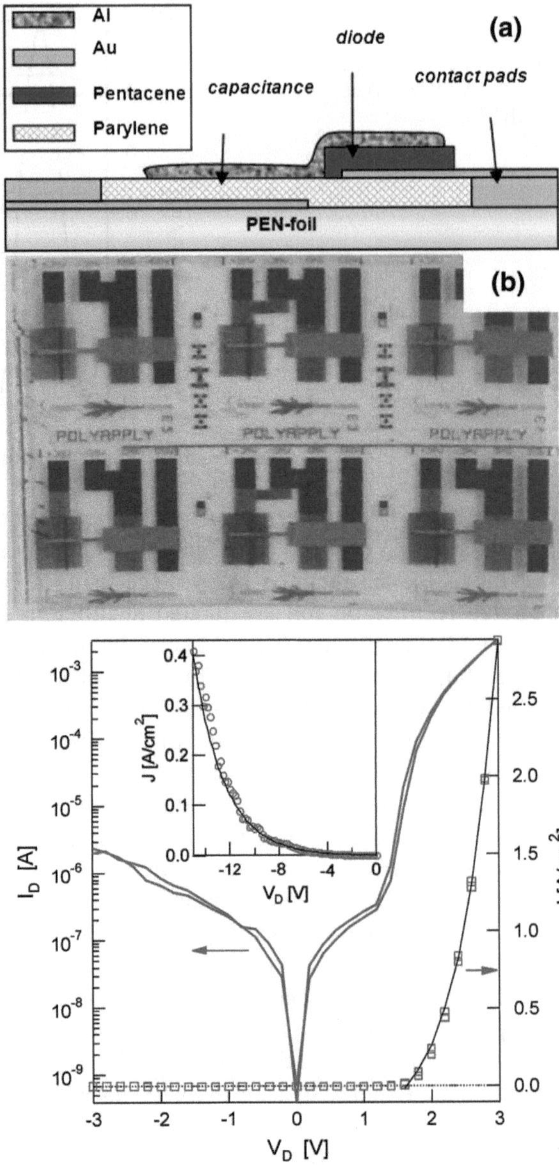

Fig. 7.11 Integrated rectifier on foil; (**a**) schematic cross-section; (**b**) photograph of the device; (**c**) I–V characteristic of an organic pentacene diode (area = 500 × 200 μm). The inset shows the diode leakage current density between 0 and −15 V. Reprinted with permission from [4]. Copyright 2008 American Institute of Physics

achieved tags by lamination of the foils, whereby electrically conductive glue is used to interconnect the different contacts of the individual foils.

The reader setup conforms to the ECMA-356 standard for "RF Interface Test Methods". It comprises a field generating antenna and two parallel sense coils, which are matched to cancel the emitted field [5–7]. By this method, only the signal sent by the RFID tag is read out at the reader side. The detected signal is

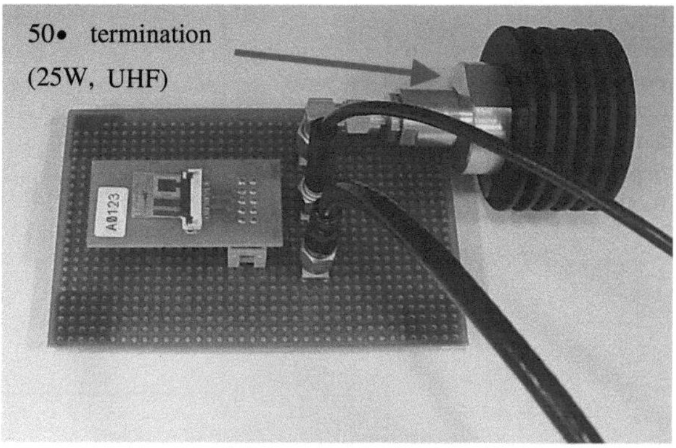

Fig. 7.12 Photograph of the high frequency rectification measurement setup for integrated rectifier

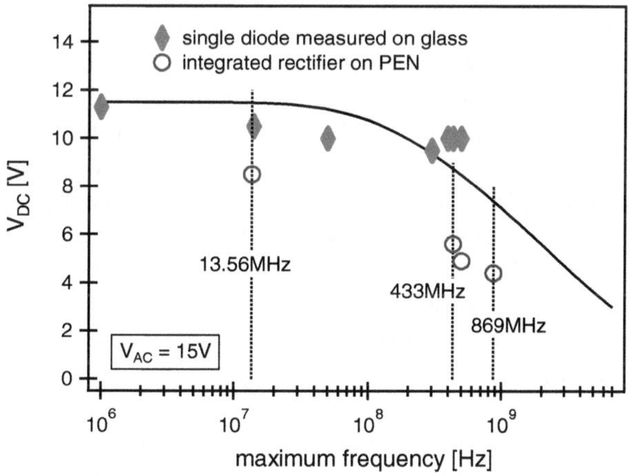

Fig. 7.13 Frequency versus rectified DC voltage for an AC amplitude of 15 V. *Diamond*: measurement discrete diode; *circle*: measurement integrated rectifier; *solid line*: theoretical frequency limit for mobility = 0.15 cm^2/Vs, $d_{pentacene}$ = 160 nm Reprinted with permission from [4]. Copyright 2008 American Institute of Physics

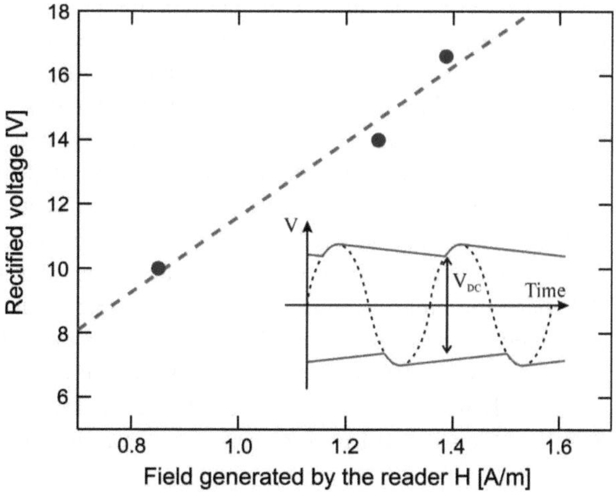

Fig. 7.14 Internal rectified voltage of a double half-wave rectifier generated in an organic RFID tag versus the 13.56 MHz magnetic field generated by the reader. (*Inset*) schematic operation of a double half-wave rectifier (see footnote 1)

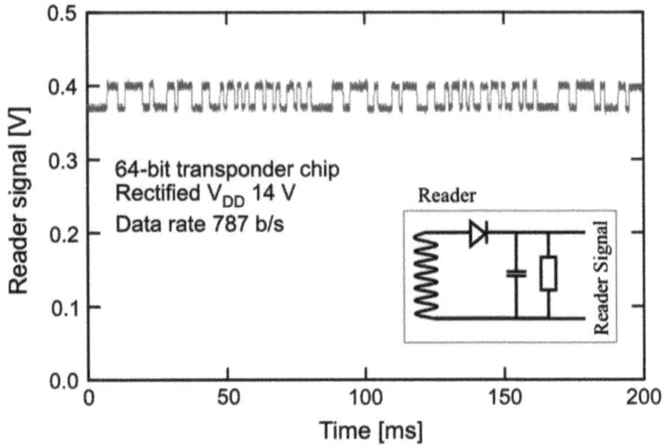

Fig. 7.15 Signal of the 64-bit RFID tag measured on the reader (unamplified reader signal). The envelope detector of the reader is depicted in the inset (see footnote 1)

then demodulated by a simple envelope detector (inset Fig. 7.15), made by a diode followed by a capacitor and a resistor, and shown on an oscilloscope.

The complete schematic for the inductively-coupled, 64-bit, organic RFID tag is depicted in Fig. 7.2. As mentioned earlier, the tags are composed of four flexible foils, with the following components: an inductor coil, a capacitor, a rectifier and a transponder. The coil is made from etched copper on foil, and was manufactured by Hueck Folien GmbH. The HF-capacitor consists of a metal–insulator-metal stack (MIM stack), processed on a 200 μm thick flexible polyethylene naphthalate (PEN) foil (Teonex Q65A, Dupont Teijin Films). The insulator material used for the capacitor is Parylene diX SR. As transponder foil, we used the 64-bit code generator described in Sect. 7.3.2 of this chapter.

The rectifier has to be able to rectify at the base carrier frequency of 13.56 MHz and to deliver a minimal DC supply voltage of 14 V, sufficient to drive the transponder. In order to achieve this, a more complex rectifier circuit has been implemented, more specifically a double half-wave rectifier [4]. A double half-wave rectifier circuit consists of two single half-wave rectifiers connected between the same nodes, as shown in the schematic in Fig. 7.2. Both single half-wave rectifiers rectify the AC input voltage: one rectifies the upper cycles of the AC input voltage, the other single half-wave rectifier rectifies the lower cycles of the input voltage. The power and the ground voltage for the digital logic of the RFID tag are taken between both rectified signals as depicted in the inset of Fig. 7.14. This figure also plots the rectified voltage obtained in DC-load modulation versus the field generated by the reader. As can be seen in the graph, 14 V rectified voltage is obtained at a 13.56 MHz electromagnetic field of 1.26 A/m. The ISO 14443 standard states that RFID tags should be operational at a minimum required RF magnetic field strength of 1.5 A/m. The double half-wave rectifier circuit presented here therefore satisfies this ISO norm. After extrapolation of the measurement data, a DC voltage of 17.4 V can be obtained at a field of 1.5 A/m. If a single half-wave rectifier was used, the rectified voltage would be limited to 8–9 V, which is too low for current organic technology.

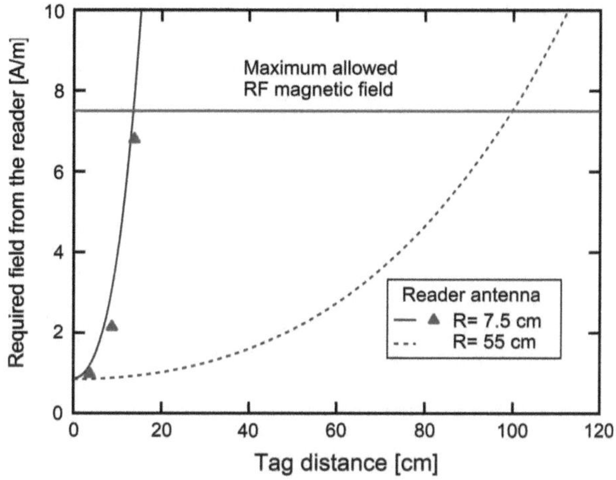

Fig. 7.16 Calculation (*continuous line*) and experimentally obtained data (*symbols*) of the required RF magnetic field at the reader side as a function of the tag distance in order to generate the required RF magnetic field to operate an 8-bit organic RFID tag

The rectified voltage of 14 V supplies the transponder chip, which sends the code to the modulation transistor. The signal sent from the fully integrated, plastic tag is received by the reader and subsequently visualized using a simple envelope detector (see inset Fig. 7.15) without amplification. The signal measured at the reader side is depicted in Fig. 7.15. This shows the fully functional, 64-bit RFID tag using an inductively-coupled 13.56 MHz RFID configuration with a data rate of 787 b/s.

Two of the reader standards at 13.56 MHz base carrier frequency are the proximity (ISO 14443) and vicinity readers (ISO 15693). The main difference between them is the coil radius, being 7.5 cm for the proximity reader and 55 cm for the vicinity reader. This results in a maximum readout distance of 10 cm for the proximity and 1 m for the vicinity reader. As mentioned earlier, the standard (ISO 14443) states also that the tag should be operational at an RF magnetic field of 1.5 A/m, which is significantly lower than the maximum allowed RF magnetic field of 7.5 A/m. In our case, the required field for an 8-bit organic RFID tag was 0.97 A/m. One can calculate the required magnetic field at the reader antenna centre in order to obtain the required field at the tags antenna to operate the tag, as a function of the distance between the tag and the reader. This calculation is depicted in Fig. 7.16. The dots in this graph show the experimental data at distances of 3.75, 8.75 and 13.75 cm with respect to the field generating antenna. This graph shows that it is possible to energize the 8-bit organic RFID tag at maximum readout distances for proximity readers below the maximum allowed RF magnetic field.

7.3.5 Sensor Integration

One of the advantages of organic technology on foil is the ease with which heterogeneous integration of a broad range of sensor technologies can be realized.

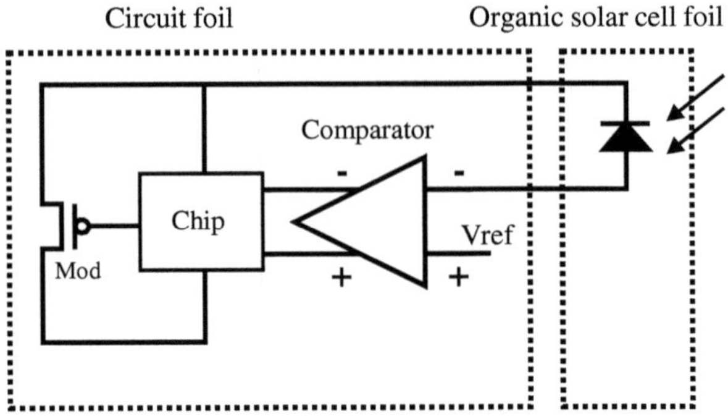

Fig. 7.17 Block diagram of the 4-bit code generator with integrated light sensor

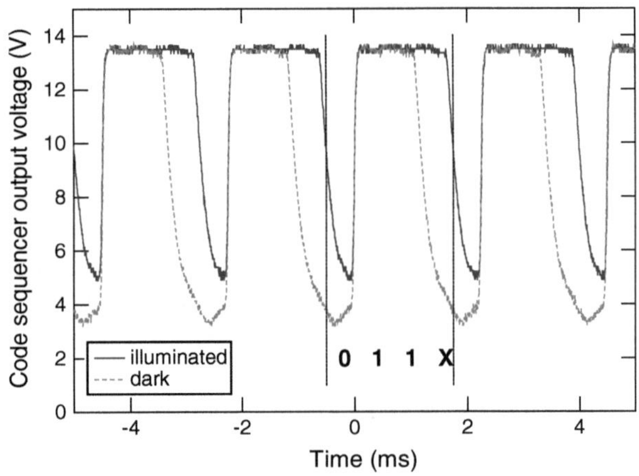

Fig. 7.18 Measurement of the code 011x generated by the 4-bit code sequencer logic, whereby x = 1 when the V_{oc} of the organic solar cell surpasses 0.2 V

As examples of possible sensor technologies on foil can be mentioned: temperature profiling, pressure and weight distribution mapping, light, pH, gas and humidity sensing. The electrical signal of the sensor needs to be read-out by the organic circuitry, which requires an analog-to-digital convertor (ADC) in most cases. A single comparator is the simplest case of ADC, providing one bit information. For more complex organic ADC architectures on foil, we refer, for instance, to [18, 19].

As a first step towards the integration of sensor functionality inside the organic RFID tags, we added a simple light sensor and a comparator inside the code sequencer of an RFID block (see Fig. 7.17). The code sequencer generates the code 011x, whereby x = 1 when the comparator measures an open circuit voltage at the organic solar cell that exceeds a set value, and x = 0 otherwise. In this embodiment, we selected an open circuit voltage of 200 mV, corresponding to a V_{ref} of 14.8 V.

Figure 7.18 shows the measured output of this code generator under both illuminated (blue curve) and dark (red curve) circumstances.

The output of this sensor RFID code sequencer comprises only one bit of sensor information (i.e. ambient light on/off). This is still a simple example, but already enables specific applications (e.g. book open at a specific page).

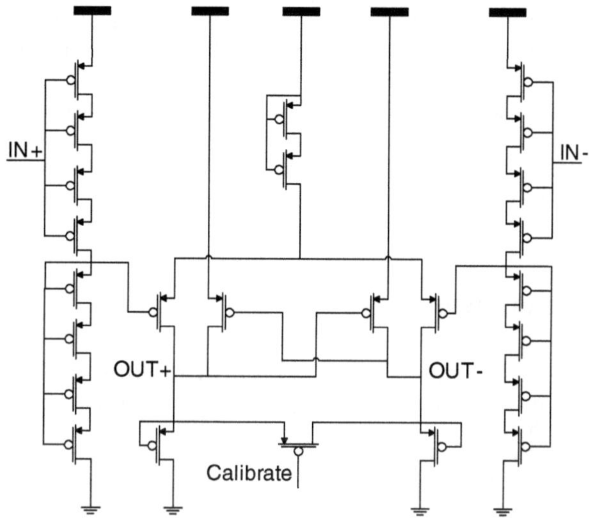

Fig. 7.19 Schematic of the implementation of the comparator block

The most crucial block in the schematic of Fig. 7.17 is the comparator. It must be designed such that it accurately switches at a given reference voltage. Figure 7.19 shows the schematic of the comparator that drives the code sequencer. During the calibration phase, all currents and voltages settle such that equal outputs are obtained, independent of the input voltages. Subsequently, when the calibration transistor is deactivated, a difference in output voltages is obtained depending on the difference in the input voltages and independent from the threshold voltage differences in the circuit.

7.4 Conclusions and Technology Outlook

The first section of this chapter has presented the roadmap for organic RFID tags, with a focus on EPC-standards for item-level tagging applications. In the next section, we have shown that some of these specifications have already been met, e.g. 50 kb/s bit rate of label to interrogator, 64 bits of hardwired memory, maximum power of interrogator and a base carrier frequency of 13.56 MHz.

Amongst the specifications that have not been achieved yet, the most important are programmable memory-arrays and an anti-collision protocol. Both specs require a bi-directional communication between the label and interrogator, which is an active research topic just now. Bi-directional tags will require a more complex design of the transponder chip and subsequently designs that are more robust against parameter variations of the technology. Circuit robustness can be improved by either implementing the design in a complementary technology [20, 21] or in a dual-VT unipolar technology [22, 23]. Furthermore, a bi-directional communication also requires a clock that is derived from the HF field. One of the potential candidates to solve this problem could be the use of higher mobility semiconductors, such as amorphous metal oxides, for the thin-film RFID technology.

Acknowledgments This work was performed in collaboration between IMEC and TNO in the frame of the HOLST Centre. Part of it has been supported by the EC-funded IP POLYAPPLY (IST #507143) and FP7 project ORICLA (ICT-247798). The authors also thank Klaus Schmidegg of Hueck Folien GmbH for the antenna foils.

References

1. Marsman AW, Hart CM, Gelinck GH, Geuns TCT, de Leeuw DM (2004) Doped polyaniline polymer fuses: electrically programmable read-only-memory elements. J Mat Res 19: 2057–2060
2. Cantatore E, Geuns TCT, Gelinck GH, van Veenendaal E, Gruijthuijsen AFA, Schrijnemakers L, Drews S, de Leeuw DM (2007) A 13.56 MHz RFID system based on organic transponders. IEEE J Solid-State Circ 42(1):84–92
3. Ullmann A, Böhm M, Krumm J, Fix W (2007) Polymer multi-bit RFID transponder. International conference on organic electronics (ICOE) 4–7 June 2007, Abstract 53, Eindhoven (The Netherlands)
4. Myny K, Steudel S, Vicca P, Genoe J, Heremans P (2008) An integrated double half-wave organic Schottky diode rectifier on foil operating at 13.56 MHz. Appl Phys Lett 93:093305
5. Myny K, Steudel S, Vicca P, Beenhakkers MJ, van Aerle NAJM, Gelinck GH, Genoe J, Dehaene W, Heremans P (2009) Plastic circuits and tags for 13.56 MHz radio-frequency communication. Solid State Electron 53(12):1220–1226
6. Myny K, Van Winckel S, Steudel S, Vicca P, De Jonge S, Beenhakkers MJ, Sele CW, van Aerle NAJM, Gelinck GH, Genoe J, Heremans P (2008) An inductive coupled 64 bit organic RFID tag operating at 13.56 MHz with a data rate of 787 b/s. IEEE international solid-state circuits conference—digest of technical papers, session 15.3, San Francisco, pp 282–283
7. Myny K, Beenhakkers MJ, van Aerle NAJM, Gelinck GH, Genoe J, Dehaene W, Heremans P (2009) A 128 bit organic RFID transponder chip, including Manchester encoding and ALOHA anti-collision protocol, operating with a data rate of 1529b/s. IEEE international solid-state circuits conference—digest of technical papers, San Francisco, pp 11.6
8. Myny K, Steudel S, Smout S, Vicca P, Furthner F, van der Putten B, Tripathi AK, Gelinck GH, Genoe J, Dehaene W, Heremans P (2010) Organic RFID transponder chip with data rate compatible with electronic product coding. Org Electron 11:1176–1179
9. Huitema HEA (2008) Rollable displays: the start of a new mobile device generation. 7th annual flexible electronics and displays conference USDC, Phoenix, Arizona, USA
10. Rolin C, Steudel S, Vicca P, Genoe J, Heremans P (2009) Functional pentacene thin films grown by in-line organic vapor phase speeds above 2 m/min. Appl Phys Exp 2:086503
11. Finkenzeller K (2002) RFID handbook, Edition 1. Wiley, New York, p 114, Chap. 5
12. de Leeuw D (1999) Identification transponder. U.S. patent WO99/30432, 17 June 1999
13. Tanase C, Meijer EJ, Blom PWM, de Leeuw DM (2003) Unifaction of the hole transport in polymeric field-effect transistors and light-emitting diodes. Phys Rev Lett 91:216601
14. De Vusser S, Steudel S, Myny K, Genoe J, Heremans P (2005) Mat Res Soc Symp Proc 870E, H1.4.1-H1.4.6
15. Steudel S, Myny K, Arkhipov V, Deibel C, De Vusser S, Genoe J, Heremans P (2005) Nat Mater 4:597–600
16. Zhu ZT, Mason JT, Dieckmann R, Malliaras GG (2002) Appl Phys Lett 81:24
17. Pope M, Swenberg CE (1999) Electronic processes in organic crystals and polymers, 2nd edn. Oxford University Press, New York, vol 1, p 663, Chap. 6

18. Marien H, Steyaert MSJ, van Veenendaal E, Heremans P (2011) A fully integrated ΔΣ ADC in organic thin-film transistor technology on flexible plastic foil. IEEE J Solid-State Circ 46(1):276–284
19. Marien H, Steyaert M, van Aerle N, Heremans P (2010) An analog organic first-order CT ΔΣ ADC on a flexible plastic substrate with 26.5 dB precision. IEEE international solid-state circuits conference—digest of technical papers, pp 136–137
20. Bode D, Rolin C, Schols S, Debucquoy M, Steudel S, Gelinck GH, Genoe J, Heremans P (2010) Noise margin analysis for organic thin film complementary technology. IEEE Trans Electron Devices 57(1):201–208
21. Blache R, Krumm J, Fix W (2009) Organic CMOS circuits for RFID applications. IEEE international solid-state circuits conference—digest of technical papers, pp 208–209
22. Myny K, Beenhakkers MJ, van Aerle NAJM, Gelinck GH, Genoe J, Dehaene W, Heremans P (2011) Unipolar organic transistor circuits made robust by dual-gate technology. IEEE J Solid-State Circ 46(5):1223–1230
23. Myny K, Beenhakkers MJ, van Aerle NAJM, Gelinck GH, Genoe J, Dehaene W, Heremans P (2010) Robust digital design in organic electronics by dual-gate technology. IEEE international solid-state circuits conference—digest of technical papers, pp 140–141
24. Steudel S, De Vusser S, Myny K, Lenes M, Genoe J, Heremans P (2006) Comparison of organic diode structures regarding high-frequency rectification behavior in radio-frequency identification tags. J Appl Phys 99:114599

Chapter 8
Printed Organic Chemical Sensors and Sensor Systems

Vivek Subramanian, Josephine Chang and Frank Liao

Abstract Printed and organic electronics has tremendous potential for the realization of new classes of very low-cost, ubiquitously deployable chemical sensors. The ability to cheaply integrate diverse materials through printing of appropriately formulated inks offers the possibility to realize highly integrated electronic nose sensors for such diverse applications as product quality checking, environmental monitoring, and other consumer-focused sensing applications. We review the state of the art in printed organic electronic sensors, discuss the major issues to be resolved, and identify potential pathways to success for this dynamic and rapidly emerging field.

Keywords Organic transistors · Sensors · Electronic nose

8.1 Introduction

Printed electronics has received a great deal of attention in recent years for a range of applications. Printing potentially enables a dramatic reduction in cost per unit area of fabrication versus conventional semiconductor and display fabrication techniques; as a result, there has been tremendous interest in deploying printed electronics for such applications as displays [1] and RFID tags [2]. The attractiveness of printing for displays is intuitively obvious; given the large areas and low functional densities of displays, printing potentially provides an attractive means of realizing large area displays at low-cost. Additionally, the good

V. Subramanian (✉) · J. Chang · F. Liao
Department of Electrical Engineering and Computer Sciences, University of California, 513 Sutardja Dai Hall, #1764 Berkeley, CA 94720-1764, USA
e-mail: viveks@eecs.berkeley.edu

E. Cantatore (ed.), *Applications of Organic and Printed Electronics*,
Integrated Circuits and Systems, DOI: 10.1007/978-1-4614-3160-2_8,

compatibility of printing with flexible substrates may potentially enable the realization of flexible displays, which are obviously attractive for a range of consumer applications. The viability of printed electronics for RFID isn't quite as obvious, but is enabled by the fact that in modern HF RFID systems, the cost of the overall tag is dominated by the cost of the antenna and the cost of attaching a silicon chip to the antenna or to an interposer. The use of printing potentially allows for the reduction in this latter cost; as a result, there is tremendous interest in the use of printing to realize HF RFID tags.

In addition to offering advantages in terms of cost per unit area and compatibility with flexible substrates, printing has another major advantage as well, when compared to conventional semiconductor fabrication techniques. Printing techniques in general and inkjet printing in particular, offer very easy integration of different materials on the same substrate. In fact, this is performed routinely in graphic arts; different color inks are printed on the same substrate to realize a full color spectrum printout. From the perspective of printed electronics, this offers a unique opportunity; for applications requiring integration of multiple, potentially incompatible materials, to be deposited on the same substrate, printing may offer a convenient pathway for realizing integrated systems incorporating all of these materials on the same substrate. One such application is printed chemical sensors; this shall be the focus of this chapter.

8.1.1 Printed Chemical Sensors: Motivation

Sensors represent an area of tremendous growth for the semiconductor industry. Industry-wide, there has been tremendous interest in deploying sensors for product quality monitoring, building and infrastructure monitoring, and various security/threat assessment monitoring applications. In general, a typical sensor consists of a sensing element, plus circuitry to interpret, analyze, and output a representative indicator of the sensor response [3]. In many chemical sensor systems, the cost of the sensing element dominates the cost of the overall sensor. This is the case for two reasons. First, the sensing element often incorporates materials that are different from those deployed in a typical semiconductor process flow, and thus have to be integrated on top of or besides a conventional silicon platform. Second, the physical size of the sensing element is often large and does not scale, unlike the processing circuitry, which generally obeys silicon scaling rules to a reasonable degree.

Given these facts, it is relatively straightforward to postulate that printed sensors are an attractive application of printed electronics. First, printing allows the easy integration of different materials on the same substrate. For example, the sensor could potentially be printed over an existing CMOS chip or on the same board as an attached CMOS chip. Indeed, printing allows for the relatively easy integration of potentially mutually incompatible materials, and thus opens up the possibility of implementing fairly sophisticated sensors that are either impossible or economically not feasible in existing silicon processes. Second, given the cost per unit area advantages of printing, this sensor may potentially be implemented at relatively low incremental cost over the silicon processing circuitry itself, thus allowing for a reduction in overall sensor system cost. Third, the compatibility with flexible substrates potentially enables the deployment of sensor systems in attractive form factors that can easily be deployed into their intended applications. For example, these sensors may be deployed into food or pharmaceutical packaging without significantly perturbing the packaging or assembly flow.

Overall, therefore, there is tremendous interest in the development and deployment of printed sensors. Within this chapter, therefore, an overview of the state of the art of printed sensors is provided, with a focus on printed organic sensors, since these represent the most widely described printed sensor technology to date. After introducing the motivations for considering printed organic sensors, we broadly review the general classes of existing organic sensor systems. Next, we focus on the primary intent of this chapter, namely a discussion of organic electronic nose sensors. We review the state of the art in organic electronic noses, and discuss the sensing mechanisms of the same. Based on these, we review the major physical effects determining operation of organic sensors, and finally, end by discussing system level issues that remain to be solved to realize integrated printed organic chemical sensing systems.

8.1.2 Organic Chemical Sensors: Motivation

Organic electronics has been an area of active research for several years. One of the major observed disadvantages of organic semiconductors and organic conductors has been their relatively poor environmental stability [4]. Organic semiconductors are known to show substantial drift upon exposure to air, moisture, and indeed, to a range of chemical species in their ambient. While this is a disadvantage for general organic electronics applications, this is a potential advantage for their use in sensors. Since organic semiconductor materials are highly sensitive to contaminants or chemical species in their operational environment, they can potentially be used as sensing elements in a gas or chemical sensor [5]. Unfortunately, the response of most organic semiconductors to external chemical species is highly non-specific. The same material may show highly similar responses to air, to moisture, to oxygen, or to any large number of chemicals in its environment. As a result, on their own, organic semiconductors in principle are not particularly attractive as sensing elements.

There are two potential and non-exclusive solutions to this problem. First, by exploiting the power of synthetic chemistry, it may be possible to functionalize organic semiconductors such that they show more specific responses to specific analytes [6]. This is in many ways a unique advantage of organic semiconductors. Since they are relatively easily functionalized with a diverse range of side and end moieties through the power of synthetic organic chemistry, it is potentially

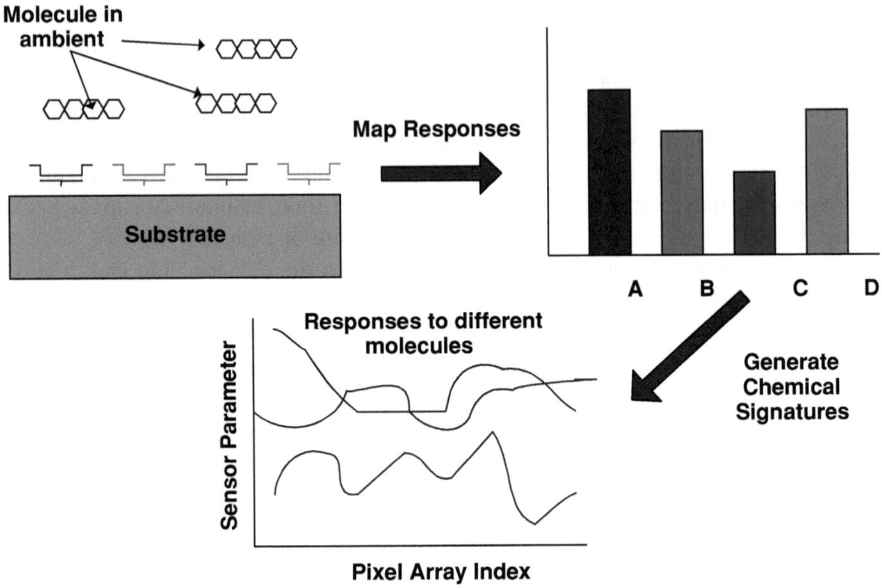

Fig. 8.1 Electronic nose conceptual basis. © 2008 IEEE. Reprinted, with permission, from [33]

relatively straightforward to optimize the response of an organic semiconductor sensing element for a specific sensing application.

The second solution to this problem, which can be used on its own or in combination with the synthetic approach above, is to use an electronic nose architecture [7]. In such an architecture, an array of individually non-specific or semi-specific sensors is connected to a pattern matching circuit. Individual sensing elements are not particularly specific, i.e., they may each respond in similar ways to a range of different analytes. However, by exploiting a sufficiently large array, it may be possible to discern a unique signature from the overall array that is highly specific. This is called an electronic nose since it is somewhat analogous to the means by which biological noses work. This is conceptually illustrated in Fig. 8.1.

As conceptualized in Fig. 8.1, the generalized idea behind electronic noses is to use an array of sensors that are non-specific [8]. Transduction elements are used to capture responses from these individual sensing elements. The set of responses for various known analytes to be detected are measured and tabulated. Upon future exposure to an unknown analyte, the measured responses are compared to the established tables and used to identify the nature of the unknown analyte.

8.2 Electronic Noses: Technology Overview

Electronic sensors for detection and identification of vapors have existed for several years. Persuade and Dodd proposed an array of sensor transducers integrated together into an intelligent gas sensor system for odor recognition in 1982

[9], and the term electronic nose was later coined by Gardner [10]. Various demonstrations of electronic nose systems have occurred since then, resulting in the development of robust systems with very good sensitivity and discriminatory power. Various demonstrations of the abilities of electronic noses have been made, including distinguishing between different vintages of wine [8] and coffee beans from different countries [11, 34].

The first generalized vapor sensing instrument to be commercially available, the "Sniffer" by Bacharach, Inc. in the 1960s, was equipped with only one sensing element, and so wasn't a true electronic nose. Sensor array-based commercial systems first became available in the mid-1990s. Early such instruments include the MOSES, the IIT Electronic nose, and the Applied Sensor 3300. Today, bench top systems are available for laboratory use, while portable systems, some including hand held components, are available for field applications [12]. In 1998, Gardner and Bartlett predicted that "the next decade should see the cost of electron noses fall dramatically and so they will be used not only in industry but also in everyday life" [10]. So far, however, this price drop has not materialized, and commercial prices for electronic noses today remain at the same level as they were in 1998.

In an archetypal electronic nose, an array of individual sensing elements is connected to a pattern matching circuit that looks for unique signatures. The potential advantage of such a system is that a wide range of analytes can potentially be detected, discriminated, and discerned using a single sensor system, making for a highly versatile sensing platform [11].

Most commercial electronic noses consists of individually fabricated sensors that are interconnected by wires or on a printed circuit board. The cost of this integration has resulted in electronic noses being rather expensive and limited to applications where cost is not a limiting factor. Indeed, as a result, electronic noses have not in general been deployed into ubiquitous sensing applications. Additionally, given the complexity of signature detection and matching, the operation of electronic noses is complicated, and the cost of developing and implementing detection algorithms has been substantial.

To solve these problems, in recent years there has been interest in two main modifications to the archetypal electronic nose solution discussed above. First, rather than attempting to develop electronic noses that are usable for a broad range of applications, the focus has shifted to developing electronic noses that are more targeted at detecting and discriminating a small range of analytes, with the electronic nose architecture being primarily used to eliminate false positives and discriminate between specific problematic analogues [7]. This dramatically simplifies the detection strategies that must be used, potentially reducing overall system cost. And, importantly for this chapter, a second focal point in recent electronic nose research has been to use printing techniques to enable the relatively straightforward implementation of integrated electronic noses, where ranges of printed organic (semi)conductors are deployed as arrays of sensing elements [6]. The use of printing, as discussed previously, enables a reduction in overall system cost and thus, the combination of these two strategies potentially allows for the realization of robust low-cost sensing systems with good immunity to false positives, offering good specificity and sensitivity.

The heart of an electronic nose is its array of gas sensors. Each sensing element is a transducer capable of converting chemical interactions with ambient vapors into an electrical signal that in some way corresponds to the vapor constituents. Transduction mechanisms can vary widely, from conductometric to capacitive, potentiometric, calorimetric, gravimetric, optical, and amperometric. Gas sensor elements that have been used for commercial electronic nose systems are listed below. Typically, two-terminal chemiresistor or three-terminal chemotransistor structures are preferred due to ease of signal processing. Gravimetric devices, which respond to vapors with a mass change that is then translated into an electrical signal, have also been used in commercial electronic nose systems. These include quartz crystal microbalance (QCM) and surface acoustic wave (SAW) sensors.

Table 8.1 provides a summary of the various classes of sensing elements that have been used in commercial electronic nose systems. In the next few subsections, a review of various types of sensing elements is provided.

8.2.1 Metal Oxide Chemiresistor Sensors

Several metal oxides, including tin oxide, zinc oxide, iron oxide, and tungsten oxide, are n-type semiconductors. When used in chemiresistor structures, the conductivity of metal oxide sensors is proportional to the carrier concentration and the carrier mobility [13]. At elevated temperatures (200–500 °C), certain odorant molecules, such as hydrogen, methane, carbon monoxide, or hydrogen sulfide, can be chemically reduced by oxygen species present at the surface of the metal oxide film. This results in increased carrier concentration and reduced barrier heights at grain boundaries, which increases the effective electron mobility within the material. The most widely used such devices are based on tin oxide resistors doped with different catalytic metal additives such as platinum or palladium. P-type devices are also available, based on materials such as copper oxide, nickel oxide, and cobalt oxide, which respond to oxidizable odorants like oxygen, nitrous oxide, and chlorine. Sensitivity depends on operating temperature and device geometry, and can be modified by catalytic dopants [13].

Elevated operating temperatures are required for these devices for two main reasons. First, the redox reactions necessary for sensing are too slow at temperatures below 200 °C. Second, chemisorbed water on the metal oxide surface inhibits sensor-analyte reactions, and must be removed by heating. The result of this need for high temperatures is that power consumption of these devices is very high, and, for portable use, power consumption must be minimized by minimizing device size and insulating the sensors to prevent heat loss.

Metal oxide sensing devices based on tin oxide were first commercialized in the 1960s. Modern variants are typically deposited through sputtering and patterned

Table 8.1 Table of sensing elements that have been used in commercial electronic nose systems

Sensor Type	Fabrication	Sensitivity	Advantages	Disadvantages
Metal oxide chemiresistor	Microfabricated, sputtering	5–500 ppm	Inexpensive	Requires high temperature operation, signal drifts over time.
Conducting polymer chemiresistor	Microfabricated, electroplating, plasma chemical vapor deposition, solution processing	0.1–100 ppm	Room temperature operation	Signal drift and sensitivity to humidity
Polymer composite chemiresistor	Solution deposited	<0.1 ppm	Room temperature operation	Signal drift and sensitivity to humidity
MOSFET chemotransistor	Microfabricated	~few ppm	Integrated with CMOS	Gate is barrier to sensing
Quartz crystal microbalance (QCM)	Microfabricated, solution processing	1.0 ng mass change	Transduction mechanism is well understood	Integration is difficult
Surface acoustic wave (SAW)	Microfabricated, solution processing	1.0 ng mass change	Good selectivity and sensitivity	Integration is difficult, sensitive to humidity, drifts over time.

using conventional microfabrication techniques to minimize device size. This complicates the integration of multiple metal oxides, since it necessitates the use of sequential subtractive processing steps. Overall, these devices offer the advantages of low-cost of individual sensors, insensitivity to humidity, and an output signal which is easy to read and process. Disadvantages include high operating temperatures, signal drift over time, limited selectivity, high power consumption, and only modest levels of sensitivity (5–500 ppm). They are also vulnerable to poisoning.

8.2.2 Conducting Polymer Sensors

In conducting polymer sensors, a chemiresistor is formed using a semiconducting polymer film that is either chemically engineered or coated to selectively absorb specific odorants. They can be used independently for detection of individual vapors or in arrays for handling of complex odors. The sensor response is produced when ambient vapors absorb into the polymer, inducing changes in the conductivity of the film. Depending on interaction between the polymer and the absorbent, this process is often reversible, but hysteresis is commonly observed, particularly when the absorbent is present in high concentrations. The sensor response is correlated with the analyte concentration, and can be modified by chemically altering the polymer or by altering sensor geometry [14].

Several key advantages of conducting polymer gas sensors makes them more suitable than metal oxide devices for sensor array applications, and as a result, the first commercially available electronic noses were based on conducting polymer gas sensor arrays. These advantages include:

- rapid, reversible responses at room temperature, which helps prolong lifetime and reduce power consumption;
- a greater sensitivity to a wide range of organic vapors;
- low-cost materials;
- resilience to poisoning by many of the compounds inorganic sensors are typically inactivated by.
- and the ease with which a wide variety of materials can be synthesized due to the flexibility of organic chemistry. This enables wide selectivity.

Conceptually, chemiresistors are fairly simple. A thin film of organic semiconductor is deposited between two electrodes. Upon exposure to a species to be detected, the chemiresistor undergoes physical or chemical changes that result in a change in the conductivity of the same. This can be measured as a change in bulk resistance, which can then be used to generate a signature for pattern matching. Selectivity can be achieved using an array of individually different organic conductors.

The transductions mechanisms are diverse and poorly understood. In general, gaseous analytes can interact with organic (semi)conductors in numerous ways. They can cause the materials to swell, resulting in increased separation between the semiconducting backbones resulting in degraded conductivity. They can dope the semiconductors, resulting in altered free carrier concentration. They can alter the trap state density of the semiconductors. And finally, they can cause twisting and/or other steric effects in the semiconductor molecules, resulting in altered carrier transport characteristics.

In chemiresistors, all these effects result in altered resistance, which can be measured using appropriate circuitry. Given these diverse interactions, there is tremendous scope to use one or more of these interactions to achieve specificity. Further selectivity and specificity can be obtained, for example, by using additives or blends of materials to tune the response variations across the electronic nose array. In such chemiresistive elements, the sensitivity depends on numerous parameters. In addition to the intrinsic responsive effects discussed above, since resistance is fundamentally a bulk measurement, the thickness and surface area of the chemiresistor has a large impact on sensor performance, both in terms of the sensitivity of the same, as well as the transient behavior and response time. Since the analyte generally has to diffuse into the matrix of the semiconductor, reduced film thickness and increased surface area is desirable to maximize sensitivity and responsiveness. The resulting signal is then passed on to signal processing circuitry, which then amplifies, digitizes, and matches the sensor output to lookup tables, thus allowing for the recognition of the analyte.

The biggest disadvantage of conducing polymer sensors is their sensitivity to humidity and their susceptibility to poisoning due to irreversible binding of vapor molecules to the sensing material [15].

8.2.2.1 Polymer Composite Sensors

Polymer composite sensors are formed using mixtures of a conductive filler, usually either carbon black or polypyrrole, and a non-conductive, but chemically sensitive polymer [16]. When used in a chemiresistor structure, resistance in polymer composite films is well described by percolation theory, in which conduction occurs through the percolation of charge through sparsely connected clusters of conductive material. Upon exposure to analytes, the insulating polymer swells, causing the conductive network to be stretched, and changing the ability of charge to percolate through the thin film. The insulating polymer is chosen for its ability to selectively absorb vapors with good sensitivity and reversibility. Because these polymers do not need to be conductive, there is much larger scope for optimizing the material choice for sensing applications [17]. Carbon black composite sensors have been demonstrated which exhibit sensitivity to amine analytes at concentrations as low as 2.7 parts per billion [18].

8.2.3 MOSFET Sensors

Another type of gas sensor can be fabricated by replacing the polysilicon gate of traditional MOSFETs with a gas-sensitive material. Metals such as palladium or platinum show a work function shift when exposed to gases such as hydrogen, carbon monoxide, or hydrogen sulfide. This results in a threshold voltage shift. The main advantage of this structure is its similarity to traditional MOSFETs used in silicon electronics, allowing for easy integration with signal processing circuitry [19].

8.2.4 Gravimetric Sensors

Gravimetric odor sensors rely on mass changes to detect vapors. The basic device structure includes a piezoelectric substrate like quartz or zinc oxide coated with a film of sorbent material. Sorption of vapor molecules into the film results in a mass change, which changes the wave velocity and resonant frequency of acoustic waves propagating through the piezoelectric substrate. Quartz crystal microbalance (QCM) sensors can detect changes in the resonant frequency, while surface acoustic wave (SAW) sensors detect changes in wave velocity. Differentiation and selectivity is achieved by modifying the coating film [13].

8.2.5 Organic Transistor-Based Electronic Noses

More recently, driven by tremendous advances in the field of organic transistor research, there has been substantial interest in the use of organic transistors as sensing elements [5, 20–22]. These have significant potential advantages over conventional chemiresistor-based arrays. To understand the potential benefits of organic transistor-based sensors, it is worthwhile to review the transduction mechanisms of the same.

The use of OTFTs as gas sensors requires a sensor device structure in which a chemically sensitive, electrically active material is accessible to ambient analyte vapors. Typically, the sensing material is the organic semiconductor, though field-effect transistor sensors have also been demonstrated in which an organic gate materials or even organic gate dielectric produces the sensor response [19, 23]. Use of the semiconductor as the sensing material is advantageous, since sensing within the semiconductor enables field-effect-based amplification of the sensor response and can cause a greater variety of changes in the OTFT I–V characteristics, including V_T shifts as well as μ_{FET} and bulk resistivity [21].

An example of an OTFT gas sensor device structure appropriate for an organic semiconductor sensing material is illustrated below. This bottom-gate, bottom contact structure leaves the active material on top, where it is easily accessible to ambient vapors. Odors flowing across the surface of the OTFT can either be adsorbed onto the surface of the active material or absorbed into the film through dissolution and diffusion. Once sorbed, the analyte can alter the electrical characteristics of the semiconductor film through either mechanical or chemical interactions (Fig. 8.2).

The first step in generating a sensor response is sorption of analyte molecules into the active material. The sorption process is described by a partition coefficient, which defines the thermodynamic equilibrium between the gas and sorbent phases of the vapor. This depends on the saturated vapor pressure of the analyte and its solubility in the sensor material, and changes with temperature and pressure. It is also affected by the presence of other vapors. In an idealized system, induced sensor response is linearly related to concentration, and the partition coefficient is constant over the range of operation conditions [24]. In fact, a linear correlation between sensor response and analyte concentration is often observed at low concentrations. By building poly (pyrrole) chemiresistors on a piezoelectric microbalances, Charlesworth et al. [25] simultaneously measured the change in the mass and resistivity of the conducting polymer thin films as they were exposed to different concentrations of various volatile organic compounds. They found that the resistance response correlated linearly with mass uptake up to a sorbed mass of 5 wt %. They also investigated the kinetics of mass sorption into the films, and discovered that sorption and diffusion of vapors into this conducting polymer differed with molecular size and chemical properties of both the vapor and the organic material. Smaller vapor molecules followed Fick's law of diffusion for planar surfaces, while larger molecules displayed non-Fickian behavior. Also,

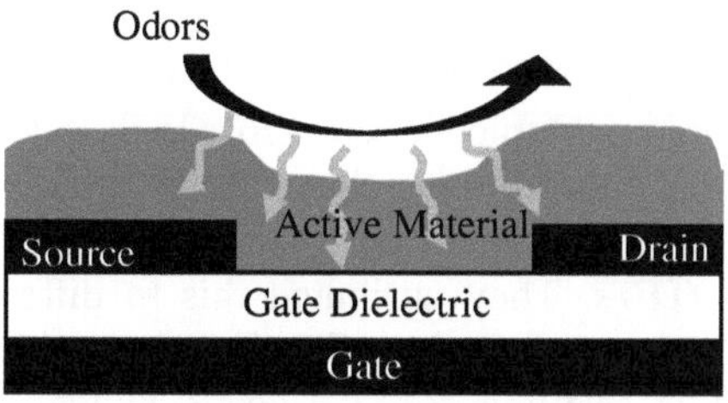

Fig. 8.2 Schematic cross-section of an OTFT gas sensor. Reprinted with permission from [6]. Copyright 2006 American institute of Physics

some vapors sorbed more readily than others, making them easier to detect at lower concentrations.

Upon removal of the vapor from the ambient, the analyte will desorb completely from the sensor film unless interactions between the analyte and sensor molecules are strong enough to either prevent desorption or kinetically hinder the process. Complete desorption allows recovery of the original electrical characteristics, while incomplete desorption results in either signal drift or poisoning of the sensor. Because strong interaction between the analyte and sensor molecule is a characteristic of both good sensitivity and an irreversible sensor response, there is a trade-off between optimizing sensor systems for good sensitivity and reusability.

The mechanisms behind the electrical response of conducting polymers to gas analytes has been the subject of much on-going investigation, and many details remain unclear [26]. Early studies on polypyrrole films suggested that a small and reversible charge transfer between the analyte and sensor molecule results in a work function shift [27]. Electrical doping and de-doping effects are commonly cited as well. For vapors with Lewis acid or base characteristics, the idea is that electron donation or withdrawal by the analyte can lead to a change in charge carrier concentration. The direction of conductivity change should in theory be predictable given the polarity of majority charge carriers and the relative magnitudes of the vapor and polymer electronegativities. Observed sensor responses do not always bear out the predictions of this theory. Many of its qualitative predictions are reliable (whether conductivity will go up or down), but a larger electronegativity difference does not consistently translate into a stronger shift in conductivity. Another proposed mechanism is the modulation of electron hopping due to changes in the dielectric property of the film when analyte vapors are present. Most models of charge transport in organic semiconductors assume that effective charge mobility is limited by the need for electrons to hop across barriers, and equations for hopping rate predict a dependence on the dielectric property of the film. A final popular mechanism cites the physical effect of vapor sorption into the film. Configurational changes due to polymer swelling by the analyte vapors may result in modified electrical behavior. In many polymers, charge flow is believed to percolate through a network of "good" connections between molecules. In this case, swelling could have a strong effect on the percolation dynamics.

The fact that most conducting polymers are used in thin-film form for sensing means that interfaces have a strong influence on both the electrical and chemical characteristics of the film and on the transduction of the sensor response. The same is true in OTFT structures, where conduction flows primarily along the surface of the organic semiconductor close to its gate dielectric interface. Tanese et al. [21] found that the electronic character of the gate dielectric, and its similarity or dissimilarity to that of the organic semiconductor, strongly influenced the sensor behavior of OTFTs. They attributed this to differences in the organic semiconductor—gate dielectric interface. Torsi et al. established that, in many analyte-sensor systems, film swelling does not occur, indicating that material incompatibilities (such as a hydrophilic/hydrophobic mismatch) prevent absorption of the analyte into the sensor film [28]. In these cases, the sensor response is elicited entirely by adsorption of the analyte at the surface or along grain boundaries in the film.

8.2.5.1 Parametric Response of OTFTs to Analytes

In chemiresistors, all of the mechanisms mentioned in the previous section will affect the one measured parameter, resistivity, in different ways, and the resulting sensor response is thus an average of all effects. In an OTFT, however, these mechanisms will affect different parameters in different ways. When no bias is applied to the gate electrode, an OTFT operates simply as a chemiresistor, and can be used to monitor the bulk conductivity of the film. This property is strongly affected by electrical doping and de-doping, and shifts in bulk conductivity typically indicate changes in the free carrier concentration in the material. When a non-zero gate bias is applied to an OTFT, the device changes from a chemiresistor to a field-effect transistor. There is some evidence that the application of gate bias can be used to enhance sensitivity or improve repeatability and recoverability of the sensor response. Electrical measurements taken in this mode can be analyzed to extract the mobility μ_{FET}, the threshold voltage V_T, and the on-current I_{on}. In organic semiconductors, charge conduction is limited mostly by hopping across potential barriers at grain boundaries. Charge trapping and de-trapping at grain boundaries is thought to modify the potential barriers at these points, modulating the carrier charge mobility. Supporting this hypothesis are studies that show that sensing in OTFTs occurs predominately at grain boundaries.

Threshold voltage (V_T) is sensitive to the work function of the material, as well as fixed charge and trap density within the organic semiconductor film. Because carrier concentration in organic semiconductors is low, band bending during accumulation mode operation typically extends through the entire thickness of the film. Evidence of this is the fact that surface potential measurements in OTFTs have been found to closely follow channel potential profiles [29]. Thus, even analytes adsorbed along the top surface of the film can affect V_T. However, the effect of charges on V_T become stronger as they move closer to the dielectric interface and new traps are more likely to capture charge as they move closer to the channel where free charge is accumulated. Because the gate dielectric interface is at the bottom of the active later, opposite the surface that is exposed to chemical vapors, V_T response is highly dependent on thickness and continuity of the active layer. In thick, continuous films, few or no vapor molecules can diffuse to the dielectric surface, and V_T response is weak. In thinner films, however, or in films with a high density of grain boundaries, vapor molecules can diffuse easily to the bottom of the organic active layer, and large changes in V_T may be observed. V_T response is expected to increase linearly with amount of vapor adsorbed, much like bulk conductivity. Thus, bulk conductivity and threshold voltage are weakly correlated. Differences in the behavior of these two indicators arise largely from the fact that V_T is also strongly dependent on the ability of the vapor molecules to diffuse effectively into and out of the organic material. Also, these two indicators show different recovery behavior, since adsorbed molecules at the surface, which affect bulk conductivity most, can desorb most quickly, while molecules at the dielectric interface, which strongly affect V_T, are more kinetically hindered.

A final parameter, the on-current of a FET device, depends on the threshold voltage, charge mobility, and carrier concentration, which is related to bulk conductivity. Thus, this indicator provides a good integration of the various indicators described above. Because this is an easy value to measure, FET current is usually favored as the ultimate indicator of electrical response. Because parameters such as μ_{FET} and V_T require mathematical extraction that is difficult to implement using analog circuitry, on-current may be the only parameter that can be monitored practically if the complexity in signal processing circuitry is to be minimized, as is desired in low-cost applications. However, because drive current is sensitive to so many factors, it is difficult to decouple the exact effects being observed. Therefore, it is useful to monitor at least one other indicator, and preferably two, along with drive current in order to shed light on the mechanisms behind the response.

8.2.5.2 Typical Sensor Responses

A typical OTFT sensor response is illustrated below. The saturation transfer curve of a poly hexyl thiophene (P3HT) OTFT taken before and during exposure to 50 ppm acetic acid, and again after purging with air, reveals that acetic acid exposure causes a reversible decrease in μ_{FET} and I_{on} as well as a positive V_T shift Fig. 8.3.

The response of a typical OTFT sensor depends strongly on the analyte, the concentration of the same, and the choice of semiconductors. In the Fig. 8.4, the differing response of two different semiconductors, P3HT, and P1 (a proprietary polythiophene-based, relatively stable semiconductor) are shown, for varying concentrations of hexanol. P3HT exhibited a positive V_T shift and a degradation of extracted μ_{FET}. Interestingly, P1 exhibited a stronger I_{on} response even though its μ_{FET} response was weaker, because it did not undergo a V_T shift. This attests the diverse responses that OTFTs show upon exposure to analyte vapors. The alcohol response for both materials was found to be quickly reversible upon removal of the hexanol, indicating that the desorption of hexanol from polythiophene is not hindered by the interactions responsible for the sensor response. Notably, many inorganic sensors tend to be sensitive to alcohol poisoning, so this repeatable,

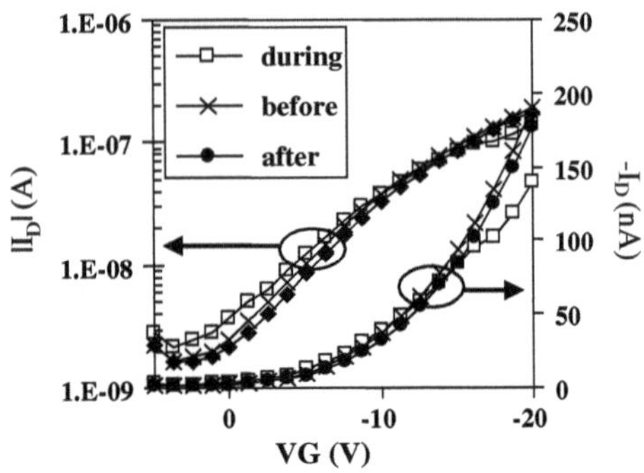

Fig. 8.3 I-V response of a P3HT sensor to acetic acid. Reprinted with permission from [6]. Copyright 2006 American institute of physics

reversible response to alcohol may be a significant advantage of organic gas sensors.

Different analytes can result in strongly different responses from the same organic semiconductors. For example, hexanethiol produced a slow degradation of both V_T and μ_{FET}, as shown in Fig. 8.5. In the presence of more than 10 ppm of hexanethiol, shifts in the both polythiophene materials continued slowly over the course of more than an hour, until the material finally became non-conductive. At analyte concentrations above 10 ppm, the rate of this shift did not correlate strongly with analyte concentration. Removal of the hexanethiol stabilized the electrical characteristics, preventing further shifts in the OTFT I–V curves, but did not reverse any shifts that had already occurred, suggesting that the thiol functional group diffuses slowly in polythiophenes, but interacts strongly with the material, preventing desorption. The opposite directions of V_T shift induced by the hexanethiol compared to hexanol in P3HT indicate that they probably introduce opposite interface or bulk charges within the active layer.

The response of OTFT sensors does not depend solely on the analyte functional group. For example, molecular weight and chain length of analytes with the same functional structure result in different responses in OTFTs. For example, Fig. 8.6 shows the response of the same sensors to a series of alcohols with varying alkane chain lengths.

This diversity of response is the key to enabling the realization of electronic nose arrays, as discussed in the next section.

8.2.5.3 Sensor Array Engineering

The gas sensor array is the core element of an electronic nose. Careful engineering of the sensitivity and specificity of individual sensing elements, as well as the collective range and discrimination of the sensor array as a whole, is thus the most crucial part of designing an electronic nose system. The ideal number of sensing elements to include in an array varies depending on the application, the specificity and range of each sensor, and the amount of correlation between sensor responses. It has been shown that the fundamental variances in analyte-sensor interactions can

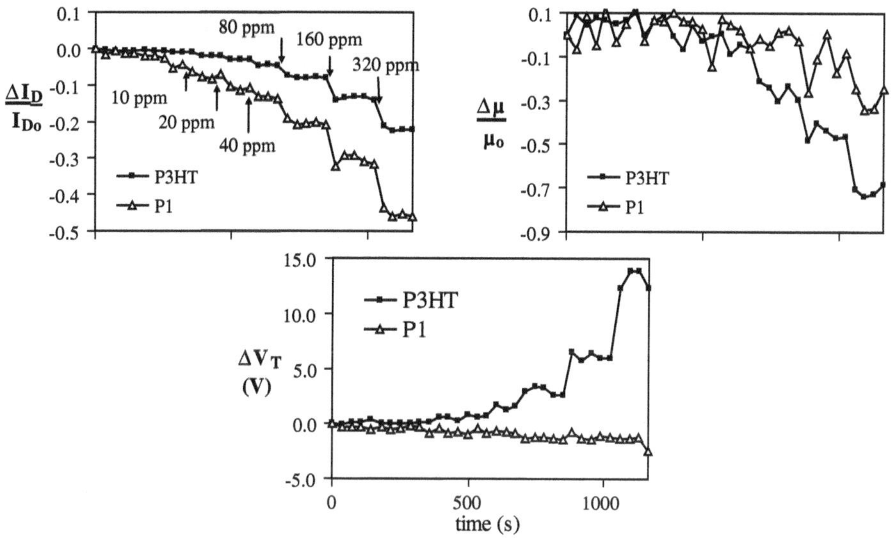

Fig. 8.4 Response of various OTFT gas sensors to hexanol exposure. Reprinted with permission from [6]. Copyright 2006 American institute of physics

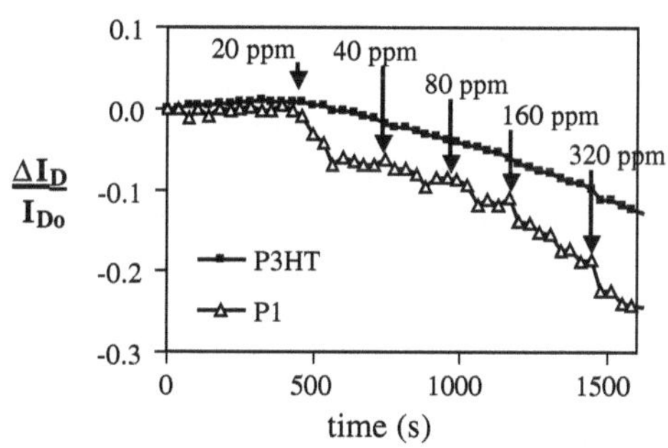

Fig. 8.5 Response of various OTFT gas sensors to hexanethiol. Reprinted with permission from [6]. Copyright 2006 American institute of physics

be summed up fairly well with as few as five molecular descriptors. Theoretically, then, five carefully chosen sensor materials should be enough for most applications. It is worth noting, however, that the human nose employs over 1,000 olfactory sensors, and a canine's nose contains as many as 100 million olfactory cells. In practice, it is difficult to find sensors whose responses are perfectly orthogonal in the odor space of interest. Rather, correlation amongst sensors will necessitate an array size much greater than five. In practical systems, the redundant information provided by overlap in sensor responses is useful for robust operation in noisy situations as a safeguard against faulty readings.

The assembly of a suitable array typically involves selection of materials adequately similar to be fabricated, implemented, and read in the same way, but chemically different enough to provide a unique set of information about incoming vapors. The introduction of different dopants or other impurities into the same

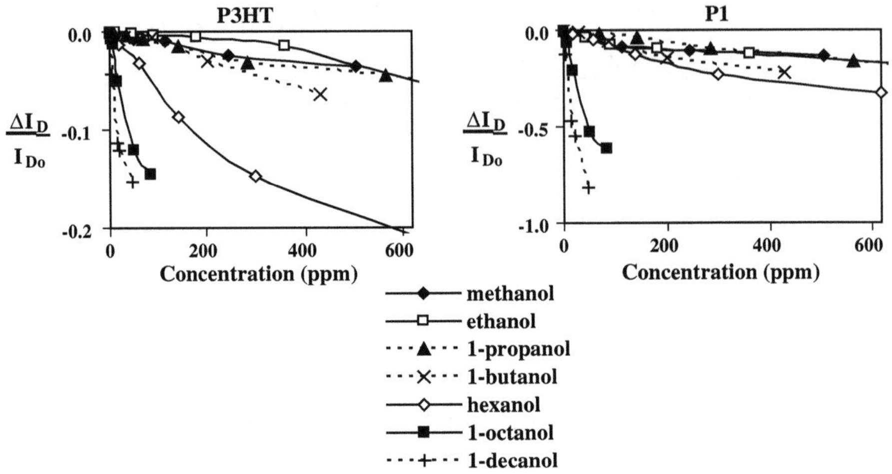

Fig. 8.6 On-current response of polythiophene OTFT gas sensors to alcohols with varying chain lengths. Reprinted with permission from [6]. Copyright 2006 American institute of physics

material is one simple method of differentiating sensor responses, though the range of discrimination achievable with this method is limited. With the advent of organic sensing materials, the use of synthetic chemistry to generate new sensor materials has become a feasible tool as well, though synthesizing new molecules and predicting their chemical properties can often be a difficult proposition.

For gas sensor arrays composed of conducting polymer chemiresistors, a variety of strategies for obtaining different sensor response have been employed. First, the starting materials for the polymer backbone can be chosen from a number of standard monomer building blocks, including single-ring heterocyclics such as pyrrole, thiophene, or aniline as well as multiring fused or unfused heterocyclics such as indole, and carbazole. These starting monomers can be further diversified by attaching side groups. Side groups can have a strong effect on molecular shape, energy band levels, charge transfer characteristics, packing, and intermolecular interactions in the final material. They have been extensively exploited to impart properties such as solubility, air stability, or self-organization.

Though synthetic organic chemistry offers almost limitless possibilities for designing molecular structures, radically new structures are often difficult to make, and successfully synthesizing a new structure can take months or even years. Furthermore, the electrical and chemical properties of even the simplest electronic organic materials are enormously complex and difficult to model or predict. Even if a new molecule can be synthesized to order, its functionality is not guaranteed, and will rarely be exactly as predicted. A better approach to tailoring the chemical sensitivity of organic sensor materials is to start with a material that is known to work as a sensor, and add different functional groups to it, preferably in locations which can modulate, but not disrupt, its ability to transport charge. Functional groups that interact strongly with specific vapors would be expected to cause a stronger electrical response to that specific vapor than to other ones, thus giving the material chemical specificity.

Extensive studies on functionalization of OTFT gas sensor material has been performed using polythiopehene. While polythiophene has disadvantages in terms of stability and moisture sensitivity, its chemistry is well understood, so it has been a workhorse for the organic semiconductor community for several decades. To design an effective library of sensing materials, a systematic method of altering the P3HT sensor response is desirable. Adding side chains can be tricky, because this often requires significant changes in synthetic route, and may disrupt packing of the polymers, resulting in non-semi-conducting films. In contrast, adding functionalized end-capping groups is a fairly reliable way of altering sensor response without destroying the polymer functionality. This approach has been successfully used by Higgins et al. [30] to make polythiophene biosensors. For use in sensors, libraries of polythiophene derivatives have been synthesized and deployed into sensor arrays [31, 32]. Typical derivatives are shown in Fig. 8.7.

By mapping out the response signatures of the various semiconductors in an array, such as the polythiophenes above, it is possible to generate a range of signatures that can be used for pattern response. A cross section of the library of reference calibration curves taken at 40 ppm analyte concentration is shown in Fig. 8.8. The functionalized polythiophenes showed significant sensor response differentiation from P3HT, as desired for electronic nose application. This attests to the power of synthetic functionalization for realization of electronic noses.

This difference in sensor responses of the functionalized P3HT materials confirms that the added functional groups do, indeed, participate in and modify the sensor response. These functional groups may modify sensor response by changing the solubility of analytes in the sensor, thus affecting the partition coefficient of the analyte-sensor system. Alternately, charge transfer or other forms of direct interaction between the functional group and the analyte are possible. It is anticipated that other functional groups, such as thiols, alkyl halides, or phosphates may also elicit distinct sensor-analyte interactions. In designing new materials, it is interesting to note that the effects on sensor response of the functional groups studied here were all fairly consistent with predictions that would be made based on the electronegativity and concentration of the functional group. The well-known chemistries of most fundamental functional groups should therefore make it possible to design a sensor array in a fairly systematic fashion.

8.2.5.4 System Level Issues for Implementation of Organic Sensors

Given the power of organic chemistry and the sensitivity of OTFT sensor arrays as discussed above, it is appropriate to end this discussion by reviewing system level issues involved with the realization of organic TFT-based sensor platforms. To perform this analysis, it is worthwhile to consider two main implementation topologies:

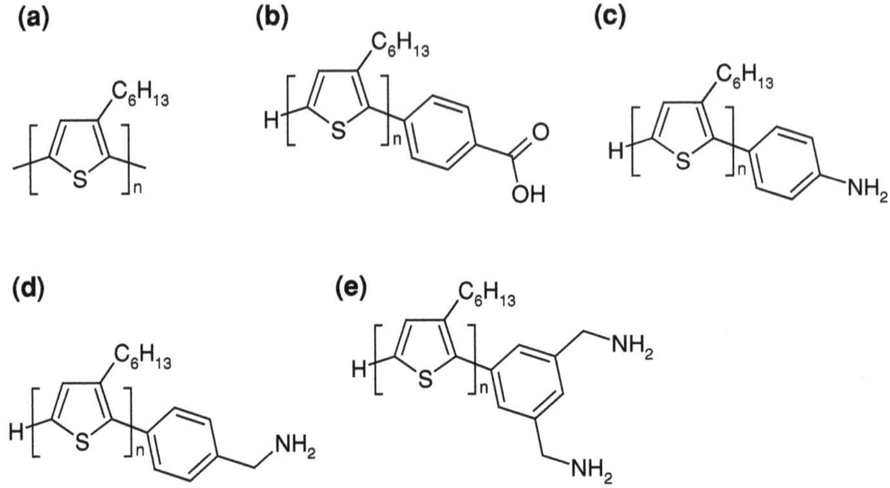

Fig. 8.7 Functionalized P3HT sensor materials. **a** regioregular poly-3-hexylthiophene (P3HT). **b** P3HT-benzoic acid. **c** P3HT-aniline. **d** P3HT-benzylamine. **e** P3HT- benzyldiamine. Reprinted with permission from [6]. Copyright 2006 American institute of physics

1. An integrated sensor system formed entirely using printed electronics, with OTFT-based sensors as well as OTFT-based signal processing circuitry
2. A hybrid sensor system, with OTFT-based sensors integrated at a board-level with silicon-based processing circuitry

The former clearly has an advantage in terms of overall integration, and potentially can lead to a lower-cost overall platform. The latter on the other hand, is easier to implement, since it can exploit the maturity of existing silicon technology.

The biggest issue associated with implementing OTFT-based sensor platforms at a system level is the complexity of dealing with the various non-idealities of OTFTs. These devices suffer from a range of drift phenomena, including drift due to bias-stress, oxygen and moisture-induced doping effects, etc [35]. To realize a viable, robust sensor platform, it is necessary to account for these drifts. The easiest way to do this is to utilize appropriate reference circuitry along with a differential sensing approach; for example, two essentially identical transistors could be deployed, one exposed to the ambient, the other protected by a passivation layer. By biasing both at the same voltage levels, elimination of bias-stress effects can be achieved using a differential architecture. Unfortunately, this does not deal with moisture and oxygen effects, however. Fortunately, in this regard, there have been substantial achievements in recent years. Recent organic semiconductors have been shown to have improved oxygen and moisture stability, and these could potentially be adapted for use in sensors, either directly as sensing elements, or in differential systems as appropriate reference devices. For example, air-stable and/or moisture-stable devices could be used to provide reference signals to identify various other

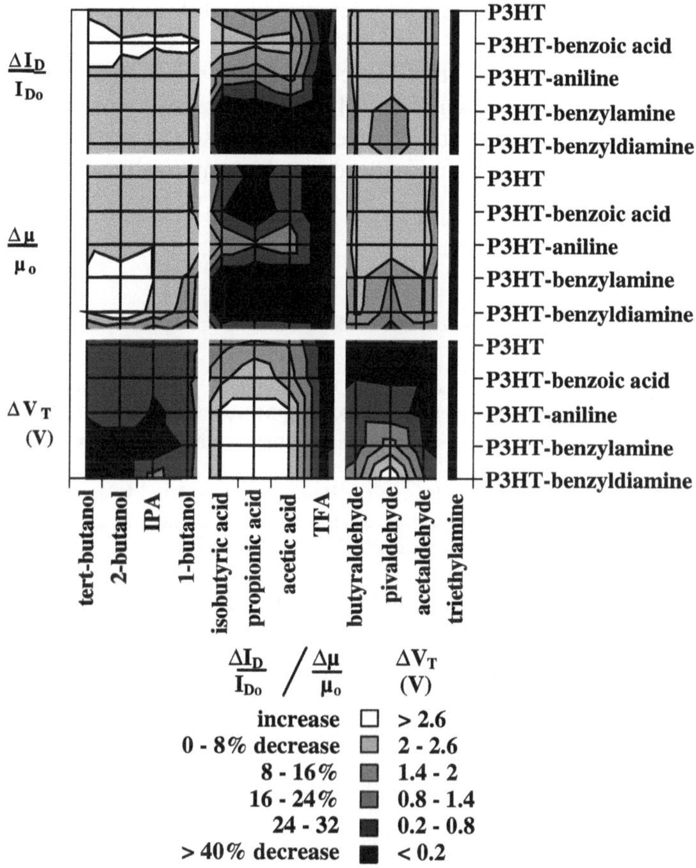

Fig. 8.8 Total sensor array response to a variety of chemicals. Reprinted with permission from [6]. Copyright 2006 American institute of physics

environmental drift effects such as temperature, etc., by providing a response to which the exposed sensors could be compared.

It is clear at this point, however, that any such sensing system will necessarily involve substantial signal processing to deal with drift and other issues. This is where implementation (1) above is so problematic. Any such sensor interrogation will necessitate the use of appropriate differential circuits, analog-to-digital converters, etc. Unfortunately, given the poor stability, matching, and reproducibility of organic transistors, a fully integrated printed organic sensor platform is not viable at this time; substantial further improvement in organic materials and device technology is required. Therefore, for the near future, it is clear that organic transistor sensors are only viable in conjunction with silicon-based processing circuitry; indeed, sensor systems based on these ideas have been reported in the literature with some success. From an economic perspective, this isn't a significant problem, since integration of silicon can likely be achieved at relatively low-cost, and it should be possible to realize fully integrated systems for less than a few tens of cents.

Overall, therefore, we see that there has been substantial progress in the development of organic gas sensors. While some issues remain, the power of synthetic chemistry makes the attractiveness of these sensor platforms substantial, and it is likely that such systems will become commercially viable in the near future.

References

1. Burns SE, Kuhn C, Jacobs K, MacKenzie JD, Ramsdale C, Arias AC, Watts J, Etchells M, Chalmers K, Devine P, Murton N, Norval S, King J, Mills J, Sirringhaus H, Friend RH (2003) Printing of polymer thin-film transistors for active-matrix-display applications. J Soc Inform Display 11(4):599–604
2. Subramanian V, Fréchet JMJ, Chang PC, Huang D, Lee JB, Molesa SE, Murphy AR, Redinger DR, Volkman SK (2005) Progress towards development of all-printed RFID tags: materials, processes, and devices. Proc IEEE 93:1330–1338
3. Gardner JW, Barlett PN (1994) A brief history of electronic noses. Sens Actuators B 18:211
4. Dimitrakopoulos C, Malenfant P (2002) Organic thin film transistors for large area electronics. Adv Mater 14:99–117
5. Torsi L, Dodabalapur A, Sabbatini L, Zambonin PG (2000) Multi-parameter gas sensors based on organic thin-film-transistors. Sens Actuators B 67:312
6. Chang JB, Liu V, Subramanian V, Sivula K, Luscombe C, Murphy AR, Liu J, Frechet JMJ (2006) Printable polythiophene gas sensor array for low-cost electronic noses. J Appl Phys 100:014506
7. Subramanian V, Lee JB, Liu V, Molesa S (2006) Printed electronic nose vapor sensors for consumer product monitoring, 2006 IEEE international solid-state circuits conference digest of technical papers, pp 1052–1059, 6–9 Feb 2006
8. Natale CD, Davide FAM, D'Amico A, Sberveglieri G, Nelli P, Faglia G, Perego C (1995) Complex chemical pattern recognition with sensor array: the discrimination of vintage years of wine. Sens Actuators B 24:801
9. Persaud K, Dodd GH (1982) Analysis of discrimination mechanisms of the mammalian olfactory system using a model nose. Nature 299:352
10. Gardner JW, Bartlett PN (1999) Electronic noses principles and applications. Oxford University Press, New York
11. Nagle HT, Gutierrez-Osuna R, Schiffman SS (1998) The how and why of electronic noses. IEEE Spectr 35:22
12. Pearce TC, Schiffman SS, Nagle HT, Gardner JW (eds) (2003) Handbook of machine olfaction electronic nose technology. Wiley-VCH, Weinheim
13. Taylor RF, Schultz JS (1996) Handbook of chemical and biological sensors. Institute of Physics Publishing, Philadelphia
14. Persaud KC (2005) Polymers for chemical sensing. Mater Today 8:38
15. Janata J, Josowicz M (2003) Conducting polymers in electronic chemical sensors. Nature 2:19
16. Albert KJ, Lewis NS, Schauer CL, Sotzing GA, Stitzel SE, Vaid TP, Walt DR (2000) Cross-reactive chemical sensor arrays. Chem Rev 100:2595
17. Severin EJ (1999) Array-based vapor sensing using conductive carbon black-polymer composite thin film detectors. Dissertation submitted to California Institute of Technology

18. Gao T, Tillman ES, Lewis NS (2005) Detection and classification of volatile organic amines and carboxylic acids using arrays of carbon black-dendrimer composite vapor detectors. Chem Mater 17:2904
19. Polk BJ, Janata J (2002) ChemFET arrays for chemical sensing microsystems IEEE sensors conference, Orlando, 5.13
20. Liao F, Chen C, Subramanian V (2005) Organic TFTs as gas sensors for electronic nose applications. Sens Actuators B 17:849
21. Tanese MC, Fine D, Dodabalapur A, Torsi L (2005) Interface and gate bias dependence responses of sensing organic thin-film transistors. Biosens Bioelectron 21:782
22. Torsi L, Tanese MC, Cioffia N, Gallazzi MC, Sabbatini L, Zambonin PG (2004) Alkoxy-substituted polyterthiophene thin-film-transistors as alcohol sensors. Sens Actuators B 98:204
23. Bäcklund TG, Österbacka R, Stubb H, Bobacka J, Ivaska A (2005) Operating principle of polymer insulator organic thin-film transistors exposed to moisture. J Appl Phys 98:074504
24. Persaud KC, Travers PJ (1997) Arrays of broad specificity films for sensing volatile chemicals. CRC Press, Inc, New York
25. Charlesworth JM, Partridge AC, Garrard N (1993) Mechanistic studies on the interactions between poly(pyrro1e) and organic vapors. J Phys Chem 97:5418
26. Adhikari B, Majumdar S (2004) Polymers in sensor applications. Prog Polym Sci 29:699
27. Topart P, Josowicz M (1992) Transient effects in the interaction between polypyrrole and methanol vapor. J Phys Chem 96:8662
28. Torsi L, Lovinger AJ, Crone B, Someya T, Dodabalapur A, Katz HE, Gelperin A (2002) Correlation between oligothiophene thin film transistor morphology and vapor response. J Phys Chem B 106:12563
29. Puntambekar KP, Pesavento PV, Frisbie CD (2003) Surface potential profiling and contact resistance measurements on operating pentacene thin-film transistors by Kelvin probe force microscopy. Appl Phys Lett 83:5539
30. Higgins SJ, Mouffouk F, Brown F, Sedghi N, Eccleston B, Reeman S (2005) Functionalized regioregular polyalkylthiophene for biosensing applications. Organic Thin-Film Electron, In: Arias AC, Tessler, Burgi L, Emerson JA (ed) Materials Research Society Symposium Proceedings 871E, Warrendale, I1.3
31. Liu J, McCullough RD (2002) End group modification of regioregular polythiophene through postpolymerization functionalization. Macromolecules 35:9882
32. Liu J, Tanaka T, Sivula K, Alivisatos AP, Fréchet JMJ (2004) Employing end-functional polythiophene to control the morphology of nanocrystal-polymer composites in hybrid solar cells. J Am Chem Soc 126:6550
33. Subramanian V, Chang JB, Fuente Vornbrock de la A, Huang DC, Jagannathan L, Liao F, Mattis B, Molesa S, Redinger DR, Soltman D, Volkman SK, Zhang Q (2008) Printed electronics for low-cost electronic systems: technology status and application development. Proc ESSDERC pp 17–24
34. Gardner JW, Shurmer HV, Tan TT (1992) Application of an electronic nose to the discrimination of coffees. Sens Actuators B 6:71
35. Crone B, Dodabalapur A, Gelperin A, Torsi L, Katz HE, Lovinger AJ, Bao Z (2001) Electronic sensing of vapors with organic transistors. Appl Phys Lett 78:3965

Index

E. Cantatore (ed.), *Applications of Organic and Printed Electronics*,
Integrated Circuits and Systems, DOI: 10.1007/978-1-4614-3160-2,

MIX
Papier aus verantwortungsvollen Quellen
Paper from responsible sources
FSC® C105338

If you have any concerns about our products,
you can contact us on
ProductSafety@springernature.com

In case Publisher is established outside the EU,
the EU authorized representative is:
Springer Nature Customer Service Center GmbH
Europaplatz 3, 69115 Heidelberg, Germany

Printed by Libri Plureos GmbH
in Hamburg, Germany